H. B. (Henry Byron) Spotton

High School Botanical Note Book

Part I.

H. B. (Henry Byron) Spotton

High School Botanical Note Book
Part I.

ISBN/EAN: 9783744777728

Printed in Europe, USA, Canada, Australia, Japan

Cover: Foto ©Paul-Georg Meister /pixelio.de

More available books at **www.hansebooks.com**

HIGH SCHOOL

BOTANICAL NOTE BOOK:

PART I.

FOR THE PRIMARY EXAMINATION.

— BY —

H. B. SPOTTON, M.A., F.L.S.

PRINCIPAL OF HARBORD STREET COLLEGIATE INSTITUTE, TORONTO.

Authorized by the Education Department of Ontario.

Price, 50 Cents

THE W. J. GAGE COMPANY (LTD.)
TORONTO.

THE EDUCATIONAL BOOK CO.
TORONTO.

Entered according to Act of Parliament of Canada, in the year of our Lord 1895, in the office of the Minister of Agriculture, by THE EDUCATIONAL BOOK CO., TORONTO.

PREFACE.

This book is designed specially to meet the wants of candidates for the Primary Examination of the Ontario Education Department. A very large number of the technical terms necessary in plant description are arranged in a systematic and convenient way, and fully defined and illustrated. In the blank schedules provision is made for entering very fully, if required, the details of structure, and for cases where additional particulars may be thought necessary blank space is provided in which such may be recorded.

Special forms of schedules for the description of Compositæ are provided, and prominence is given, in all the schedules, to drawing, which is so indispensable to good work.

In order that every facility may be afforded to the young student to acquire early the fundamental ideas of classification, an analytical table of the chief Orders represented in Canada is provided, so that as soon as a basis has been laid by the thorough examination of a few representative plants, the Orders to which they belong may be ascertained by the pupil at once. For the full classification recourse must necessarily be had later on to the Flora, which is designed to accompany the note-book.

It is hardly necessary to point out to the intelligent teacher that he must use judgment in determining the degree of fullness of description which he will demand from his pupils. Very much depends on the stage at which they have arrived; a description which would be quite satisfactory as coming from beginners might be wholly inadequate if given by advanced pupils. The schedules can of course be easily adapted for use with classes of all grades.

TABLE OF CONTENTS.

On the Management of Elementary Classes in Botany.

Practical Exercises.

Orders Prescribed for Study for the Primary Examination.

Outline of Classification.

Glossary.

Key to the Families or Orders.

Illustrative Examples of Plant Description.

Descriptive Schedules :
   Ordinary Plant Schedules.
   Composites.
   Leaf Schedules.
   Flower Schedules.

Floral Diagrams.

Index.

Blank Leaves for Notes.

ON THE MANAGEMENT OF

ELEMENTARY CLASSES IN BOTANY.

The following suggestions are offered in the hope that they may be found helpful to those who are beginning the work of teaching Botany, as well as to the young student. The writer, mindful of the difficulties and perplexities which he has himself often had to encounter, makes no apology for thus presenting what appear to him to be the chief essentials to success in this department of school work. It goes without saying that no written instructions can ever make a successful teacher where natural enthusiasm is wanting, but it is equally true that the young enthusiast may derive some benefit from the larger experience of others; and while the intelligent and active teacher will not slavishly follow the details of any method, but will be quick to avail himself of any legitimate device which will serve his purpose, still there are broad principles upon which those who have had practical experience will probably agree. In the following remarks an attempt is made to outline the course of a year's work, which it is thought will be found practicable in any High School.

When to begin Botanical Work.—A good deal might be said in favor of beginning our botanical work in the spring. At that time, when nature is awaking from the torpor of winter, and the first leaves and flowers are unfolding, it is especially delightful to ramble abroad. Then, perhaps more than at any other time, the youthful mind is attracted by the forms of the vegetable world, and is prepared to enter upon the systematic study of them with more than ordinary enthusiasm. And if it were possible to continue through the summer the botanical work begun in the spring, doubtless the most satisfactory results would be obtained. There is, however, the break caused by the long vacation, during which teacher and pupils are separated and school work generally abandoned, so that when classes are resumed in September the work of the spring has to be gone over again, with the disadvantage of having, in most cases, new pupils as well as old ones to deal with. On the whole, therefore, as the school year begins in September, and a general re-organization of classes then takes place, it seems most advantageous to begin the botanical work at that time. During September and October an abundant supply of material is available, with the advantage also of access to fruits and seeds of all kinds, as well as flowers. It is exceedingly desirable that during this period, when fresh plants can be had for examination, the botanical lessons should be frequent. If a short lesson could be given every day at this time surprising progress would be made in a few weeks. When summer has passed by, and work has to be confined to such material as has been collected for winter use, the lessons need not be given so often; probably twice a week would be found quite sufficient. Then, in the spring, when field work can be resumed, the lessons may again be increased in frequency for a time.

How to begin.—Assuming, then, that the botanical work is commenced in September, the next question to consider is how to carry on the work of the class so as to give the subject its highest educational value. Botany is essentially a science of observation. One of its very highest uses as a factor in education is that it trains the eye to habits of accuracy. But, in order to receive this benefit, it is essential that the pupil should be brought into contact with the forms which are the objects of study; that he should handle them and view them for himself; that he should by personal inspection, ascertain their habits, and by visiting their haunts learn the situations in which they flourish best. Undoubtedly, then, the first essential in giving a lesson is that every member of the class should have before him a specimen of the plant, or part of plant, which is to be the subject of the lesson. Then the teacher will direct attention to the different organs, naturally in the order of development of the organs themselves; first to the root, then to the stem, then to the leaves, and finally to the flower. In a first lesson it would not be amiss to make a superficial examination of the entire plant, rapidly and briefly discussing the nature and use of each part, but avoiding as far as posssible the use of technical terms. The chapter on the Buttercup, as given in the text-book, really affords material for several lessons. Each teacher must however, be guided by the time at his disposal and the circumstances of his class as to how much ground he will attempt to cover at one time. Some of the plants described in the text-book, as for instance, Hepatica and Marsh Marigold, will not be available in the autumn. This, however, is a matter of comparatively little consequence, as others can be substituted. In fact, after one plant, such as Buttercup, has been thoroughly understood, almost any other dicotyledonous plant may be taken up and compared with it. The order followed in the text-book is a good one, because the pupil is led by degrees from the study of floral forms in which all the parts are present but entirely disconnected, to others showing various complications and irregularities; but the judicious teacher will readily supplement the work of the text-book by the use of material which he will find in abundance everywhere about him. Let him keep in view the series of facts which it is essential that the class should know, and he may use any material which would enable the class to discover those facts from personal observation.

How to conduct a Lesson.—If the class is a large one, it will economize time to have the observations made simultaneously. Suppose the Red Maple is the subject of the lesson, which of course in this particular case must be given in the spring. The class having observed that the flowers precede the leaves, that the flower-clusters upon one set of trees differ in appearance from those upon another set of trees, and that all the trees are visited by multitudes of busy insects, let an abundant supply of both sorts of flowers be procured and taken to the class-room. Let the teacher then distribute the staminate flowers, and proceed with the observations upon them. Every pupil should have before him a blank schedule, in which he will set down the result of his observations, and it will be well for the teacher to have a large schedule, visible to the class, marked off upon the blackboard. Assuming that the pupils have been made acquainted with the common terms employed in the forms, let them all be required to examine the calyx, and to set down in the proper place the number of sepals. Then ascertain what has been thus set down. If all agree in their observations, the result may be accepted and recorded in the schedule on the blackboard. If there are variations, these must be looked into and noted, if correct. Then comes the question—"Polysepalous or Gamosepalous?" the result to be checked as before. Then—"Superior or Inferior?"—to be dealt with in like manner. To fill the last column, headed "Remarks," it will not be amiss to leave the pupils entirely to their own judgment as to what they may think worth recording. When the notes have been made, the teacher may select from them such as are most worthy, and enter these in his blackboard schedule. The corolla will next be looked for and a record made. The word "Wanting" will doubtless be written down by every one, and may then be also written on the blackboard. Then the stamens come under notice. Each will set down the number he finds, and in this case it is hardly likely that all the results will agree. Some will find five, others six, others seven. When all the results have been ascertained, the teacher should enter in his form the lowest and highest numbers, thus: 5-7, as expressing the collective result, and he should improve the opportunity here presented to caution his pupils not to infer too much from the examination of a single specimen, as variations similar to that now under notice are not uncommon. The cohesion and adhesion of the stamens will next be observed, and the "Remarks" column filled and checked as before. Pistil "Wanting" will be the next entry, and will complete the examination of the staminate blossom. The fertile flowers will then be distributed and the work carried on in the same manner, the pupils being led to find out for themselves the difference between the two kinds of flowers, and no observation on their part being considered altogether unworthy of notice, even though relatively unimportant. The significance of the invasion of the flowers by insects can now be made clear, and the pupils should be advised to observe the trees from time to time afterwards, in order to see what progress the fruit is making, to note the development of the wings, the lengthening of pedicels, and finally the sprouting of the seeds and the production of a strong new plant, all in one season. Other points, such as the shape of the leaves, comparison with other species of maple, etc., etc , may be introduced at the discretion of the teacher, but care should be taken to avoid vagueness and confusion in offering for the consideration of the pupils more than they can readily grasp, and the *relative* importance of points of structure should be distinctly brought out. For this reason a form of schedule, which will present the various features in their proper perspective and avoid giving the impression that all observations are of equal importance, is the best. It is, in the writer's judgment, a great mistake to dwell at first with any degree of minuteness upon the morphology of the various organs—to attach much importance, for instance, to the minute description of leaves. What is wanted is to get a clear apprehension of the leading characteristics of the great groups of plants, and the main facts of plant life, and anything which tends to cloud the perception of these things must be a hindrance to true progress. After typical floral forms have been examined, and some knowledge has been gained of the more comprehensive groups, then it will be proper to proceed with the study of those finer distinctions upon which depends the separation of genera and of species, and which are essential to know in order to use intelligently the classified list of the common plants of the country.

Winter Work.—As already suggested, the lessons in the fall should be as frequent as circumstances will allow, so as to complete the examination of as many typical flowers as possible. Meanwhile preparations should be going on for the winter lessons. Fruits, seeds, leaves, bulbs, tubers, cones, etc., etc., should be collected in as great variety as possible. A supply of ferns should also be laid in, neatly pressed and mounted, as these plants may be studied nearly as well in winter as in summer. Elementary microscopic work can also be just as well done in winter. Every school should now have a good compound microscope, and the teacher who can skillfully cut a few hand sections has at his command an inexhaustible source of interest and delight to his class. In all this winter work, and indeed in all botanical work, a good deal of attention should be given to *drawing*. It forms a very useful exercise, for example, to dictate or write on the blackboard, a botanical description of a leaf, and then require the class to draw the leaf so described. So, also, if a section is viewed through the microscope, a drawing of what has been observed should in all cases be demanded, as the most satisfactory way of ascertaining whether the observer has carried away the right impression; whether he has, in short, seen what he was desired to see. It will often happen, too, in the examination of minute flowers, that it becomes necessary to dissect out and exhibit separately special portions of the flower, say, for example, the pollen-masses of the milkweed, or a single stamen of the pine. The teacher should, in such a case, perform the necessary dissection; and having fixed the portion properly under the lens, pass it round for the inspection of the pupils. They may then be required to make a drawing of the object, and having thus apprehended what is necessary, may be asked to try to repeat the dissecting process for themselves.

The study of the structure and germination of seeds is another part of the work which can be very well done in winter, and many interesting and valuable lessons may be given upon these points. Seeds of different sorts should be placed upon wet flannel or blotting-paper and allowed to germinate. The whole process may then be observed in the most convenient way, dissections and comparisons being readily made at various stages.

Spring Work.—If the programme thus lightly sketched be fairly carried out, the young botanist should be very well prepared for field work in the spring. He will now put to a practical use the information he has been acquiring about the parts of plants and their modifications, and will proceed to identify and classify the flowering plants which come in his way according to the characters which he finds them to exhibit. As soon as practicable the pupils should be required to collect and bring to the class-room any wild plants whatever which they may find in flower. If their specimens are enclosed in a suitable tin box, with a light sprinkling of water, they will remain in excellent condition for several days. It is now of minor consequence whether all the members of the class are engaged upon the same plant or not; but whether they are at work upon the same or different plants, the schedules must in all cases be conscientiously filled up before they attempt to determine the name and place of the specimen. Perhaps, for a time, until all become familiar with the use of the "flora," it would be better to work upon one plant at a time. If this plan be followed, the points of structure should be observed, recorded, and checked as already described for the earlier lessons, and when the characters of the plant have thus been definitely settled, recourse must be had to the "key" which is prefixed to the flora. Full instructions are given in the book itself for the use of this "key," so that they need not be repeated here. All the teacher has to do is to accompany the class through the various questions which have to be answered, putting them, if preferred, one by one, and receiving the answers of the class in any way he may prefer, the answers in every case, of course, to be obtained from the completed schedule. If the true name of the plant is at length arrived at, this will be the best evidence that the work of observation has been accurately performed. Two or three lessons carried out in this manner will give the pupils confidence, and familiarize them with the use of the flora, after which they may be allowed to examine and determine almost any flowering plant they may meet with. The teacher will find it useful at this stage to begin a register of the practical work done by his pupils. If prizes are given, the awarding of them may be made to depend largely upon the showing of this register. Then, if there is time, the mode of *preserving and mounting specimens* for the herbarium might be explained. Apart from its botanical importance, this work has an educational value in itself, demanding, as it does, the greatest neatness and care to ensure the most successful results. Full instructions will be found at the end of the glossary.

Excursions.—The writer cannot do better than reproduce here a short account of a botanical field day, written by him for the *Educational Monthly* some time ago, in order to illustrate how such a day may be spent:—

A BOTANICAL FIELD DAY.

It is a bright Saturday morning towards the end of June—a morning to which a score of boys and girls have for some time been looking forward with a good deal of pleasant anticipation. They are juvenile botanists, members of a class formed some months ago, and having now, by the study of selected specimens, acquired some little knowledge of the structure of plants, they are, on this particular morning, to meet for a ramble; together to behave as come in their way; and then to re-assemble and compare notes, and also to determine the names of such plants as they do not already know.

The rendezvous selected is a particularly good one for botanical purposes, commanding, as it does, a variety of situations. It is an upland from which, by a gentle slope to the northward, you may descend to the reedy margin of a small lake, concealed by trees until you are close upon it. East of this lake stretches a beaver meadow of many acres, fringed and dotted with larches, and too moist to traverse in comfort at most seasons of the year, but, in this warm and leafy month of June, solid enough under foot to dispel uncomfortable fears of false steps. If, instead of descending, you skirt along the brow of the hill, to the westward you come upon open meadows, with here and there a low copse or thicket; while to the eastward are noble woods of maple and beech, succeeded farther on by pines, as the character of the soil changes. To the southward are cultivated fields and market gardens, and in the distance the glinting of the sun on a couple of church spires marks the direction of the neighboring town.

Ten o'clock is the hour of meeting, and on this occasion an exemplary punctuality is observed by everybody. As it is intended to make a day of it, lunch baskets have not been forgotten. These are left for safekeeping at a cottage close by, and then, after a brief rest in the shade of a friendly beech, the party is divided, for the day's work, into small groups, and an area roughly marked out for each. The lower grounds and the lake region, as being somewhat difficult of access, are assigned to the sturdier boys, whilst the hillside and the exploration of the woods and fields above are divided among the remainder.

It is agreed that the work of collection shall be limited to two hours, and accordingly, as the distant boom of the noon bell comes over the fields, our botanists begin to straggle in again. It is nearly one o'clock, however, before the last detachment arrives. This consists of the boys who have made their way to the eastern end of the lake and the beaver-meadow. Their appearance is hailed with a shout of admiration, for of all the collections of flowers, theirs is certainly the most imposing. They must, indeed, have hit upon a veritable botanic garden, for each of them carries a huge bouquet, made up of a profusion of Lady's Slippers and other Orchids, together with Lilies, Pitcher Plants, and beautiful pink Pyrolas. These boys are flushed with the excitement of their walk and their success; and though the condition of their lower extremities would seem to indicate that they are not altogether unacquainted with bogs, they make no reference thereto, but dwell with enthusiasm, and some degree of extravagance perhaps, on the beauties of the scene they have just left. But the others, though their collection will not vie in brilliancy with the products of the beaver-meadow, have, nevertheless, in nearly every case, something of more than ordinary interest to show. The explorers of the lake margin were fortunate enough to find a punt, by means of which a number of aquatic plants. Yellow Pond Lilies, Utricularias, the pretty white Water-Crowfoot, and the Water-Shield, were brought within their reach; and on the cool northern hill-side, trailing over the base of moss-covered stumps, specimens of the Twin-Flower—a special favorite of the great Linnæus, and named *Linnæa borealis* in his honor—were obtained, as well as Violets of various species, Woodbines, Mitchellas, etc. The open fields and fence-rows yielded St. John's-worts, Elder, Gnaphaliums of several species, a handsome Rudbeckia—the purple Cone-flower—and of course the ubiquitous Dandelion, and Mayweed, and Mullein.

But just now there are cravings which are not intellectual, cravings too urgent to be disregarded. The interest in botany is, at this moment, decidedly of a secondary nature, and when the lunch baskets are sent for, and their contents exposed to view, the gravest doubts of their sufficiency are entertained and freely expressed. The fullest kind of justice is done them, and in the course of a few minutes no vestige whatsoever remains—nothing even suggestive of them, save the shrunken wrappers, upon which some eyes are now turned with an expression almost approaching to gloom. It is suggested, and the suggestion meets with no opposition, that whatever may be the merits of botanical pursuits from an intellectual point of view, they have recommendations of a physical nature, not wholly unworthy of consideration; and it begins to dawn upon these youthful scientists, though as yet they have no clear conception of the ideal *mens sana in corpore sano*, that Botany has this decided advantage over all other school studies, that, to pursue it with efficiency, exercise of body must accompany exercise of mind. They can also comprehend that the botanical laboratory is as free as air to everyone who wishes to make use of it; that everywhere around them the lavish productions of nature are only waiting to be asked, to unfold their beauties; and that anyone who holds converse with the silent yet eloquent creations of the floral world, must become imbued with more or less of the feeling which inspired the tenderest of American poets, when he sang of the flowers as

"Teaching us by most persuasive reasons
How akin they are to human things."

But the afternoon is advancing, and important work still remains to be done. It is not enough to admire color and form; we must look a little deeper, and analyze the structure of our flowers with as much minuteness as may be suited to the capacity of the present students. In other words, we propose to turn our ramble to practical account in the way of an object-lesson, and to test the observing faculties by trying to assign to each plant its proper place in a botanical classification. A good many of the plants are recognized, without much difficulty, as being near relatives of species already examined in the class-room; the Lady's Slipper, for instance, is at once pronounced to be an Orchid; the Pitcher-Plant is immediately identified by its leaves, the Water-Crowfoot is only a white Buttercup; the few Composites in bloom at this season are referred at once to the proper family; and so with a number of others. But there are some which cannot be disposed of in this off-hand manner, and for these our "Flora" must be consulted. For convenience, it is arranged that one person shall read aloud from the manual, while the others, with specimens in hand, listen to the descriptions, and assent or dissent, as these correspond to the characters exhibited by the plant under examination, or the reverse, until finally its true place and name are revealed. These having been duly noted down, along with the date of collection and the locality, other specimens are taken up in the same way; and though it is found impossible to overtake all the plants that have been gathered, yet considerable headway is made, and even the dullest (for our class, not being an ideal one, contains dull as well as clever pupils) feel a certain degree of confidence in their ability to do a little botanical work on their own account.

The work of determination is not prolonged to weariness, and soon after three o'clock preparations are made to return home. The fatigue of the morning's walk has completely disappeared, and the youthful mind, released from the strain to which it has been subjected, unbends, and with that singular fertility of resource which causes the average juvenile to be at once the envy and the terror of his elders, immediately advances a host of topics for discussion, quite foreign to the object of the day's proceedings. Botany is for the present laid aside, and it ceases to be a matter of any consequence whatever, whether stamens are hypogynous or otherwise, or what may be the relation of the calyx to the ovary. With pleasant conversation the homeward way is beguiled, and as we separate, a hope, which is believed to be genuine, is expressed that ere long we may meet again for another Field Day.

PRACTICAL EXERCISES.

1. —Examine and record, with drawings, the modes of vernation in six different plants.
2. —Compare the leaves of Red Maple, Silver Maple, and Sugar Maple, making drawings.
3. —Compare the leaf-clusters of the White Pine, Red Pine, and Tamarack.
4. —Determine the phyllotaxis in six different plants.
5. —Make a cross-section of a cluster of the leaves of the Blue Flag, near the base. Make a drawing of the section.
6. —Examine buds of the following, with special reference to protective coverings: Lilac, Spruce, Horse-chestnut, Beech, Poplar. Make notes of what you observe. Where bud-scales are present examine their inside surfaces.
7. —Compare the climbing apparatus of the Pea with that of the Bean.
8. —Compare as to mode of growth and ramification the stems of the Apple-tree and the Pine.
9. —Make vertical sections of the eye of a Potato, an Indian Turnip, and an Onion, and make drawings of the sections.
10. —Make vertical and cross-sections of three different buds. Draw the sections.
11. —Examine the prickles of a Bramble and of a Galium. Are they hooked downwards or upwards? Of what service are they to the plants? Give reasons for your opinion.
12. —Examine the ends of shoots of the Lilac towards the close of summer. Note the replacement of the terminal bud by two lateral ones. Examine these again late in the fall.
13. —Examine tendrils of the Grape-vine and Virginia Creeper, noting any difference in their mode of action.
14. —Examine the twining stems of the Hop and the Morning Glory, noting differences.
15. —Detach bulblets from the axils of the leaves of the Tiger Lily, and plant them. Record results.
16. —Cut with a knife into the stems of an exogen and a woody endogen (Bamboo, for example). Note and account for any difference in the difficulty of cutting through the outer surface.
17. —Examine and record, with drawings, the modes of æstivation in six different flowers.
18. —Draw floral diagrams of six different flowers, and write out the formulas.
19. —Compare the *head* of the Thistle with that of the Red Clover.

20.—Detach with the point of a pencil the pollen-masses in any orchid flower, thus imitating the action of an insect. Note the downward contraction of the pollen-mass shortly after its withdrawal. What purpose is served by this contraction? Extract also the pollen-masses from a flower of Milkweed.

21.—Observe whether insects visit the flowers of any of the following: Pine, Willow, Cucumber, Maple.

22.—Make and draw sections of six different ovaries.

23.—Soak a bean in water for an hour or two, and then dissect it, exhibiting all its parts.

24.—Compare the pappus of the Dandelion with the silky hairs upon the seeds of Milkweed and of Willow-herb. Note differences of origin.

25.—Bury a bean and an acorn in moist, warm sawdust, and note any difference in the phenomena of germination.

26.—Gather a few acorns and seeds of the Red Maple and lay them away for the winter. In the spring test their germinating powers.

27.—Examine scales of green pine-cones, and also of ripe ones.

28.—Study the dehiscence of the ovary in Purslane, Shepherd's Purse, Catchfly, Columbine, Mallow, Morning Glory.

29.—Dissect out the embryos from six albuminous seeds.

30.—Observe through a good microscope, and make drawings of:—

    (a) Six different pollen-grains.
    (b) A thin slice of Elder pith.
    (c) A shred torn from the under surface of a leaf.
    (d) A similar shred from the upper surface.
    (e) A cross-section of a bit of Lilac leaf with a vein in it.
    (f) A plant-hair.
    (g) A vertical section through the tip of a rootlet.
    (h) A thin slice of Potato.
    (i) The bloom on a Cabbage-leaf.

31. Make cross-sections of the Bamboo and a branch or small stem of any of our native woods. Examine with a lens and write notes on the different appearances presented.

32.—Examine the bark of a young tree and also of an old one of the same kind. Note any differences and account for them.

33.—Examine a bit of the under side of a leaf of Sweet-brier under a good microscope. Give your opinion of the source of its colour.

34.—Examine the sticky stem of the Catchfly. What causes the stickiness? What is its probable use?

PRACTICAL EXERCISES.

35.—Examine the scurfy under surface of a leaf of the common Shepherdia. View a small portion under a good microscope and write notes on what you observe.

36.—Scrape the surface of a slice of Potato with a knife, mount the scraping, and examine with a good microscope. Add a drop or two of solution of iodine; examine again, and describe and explain the result.

37.—Try similar experiments with a Turnip, a Carrot, an Apple, a softened Pea, and write notes on the results.

38.—Study the germination of a Pea, a Windsor Bean, and a grain of Indian Corn. Write notes upon any phenomena observed. Try the effect of different temperatures on the rapidity of germination.

39.—Observe and write notes upon the different aspects presented by plants when grown in the shade and when exposed to full sunlight.

40.—Immerse a few green leaves in a bottle full of water. Invert upon a shallow dish of water without spilling. Expose the whole to strong sunlight, and examine after two or three hours. Describe and explain anything you observe.

41.—Repeat the last experiment, placing the apparatus in a dark closet. Note results.

42.—Fill about one-third of a large wide-mouthed bottle with well-soaked Peas. After three or four hours carefully remove the stopper and lower into the bottle a lighted match or taper. Note and explain results.

43.—Grow a hyacinth or a crocus in a perfectly dark cellar. Note the effect upon the colour of the leaves, and also upon that of the flowers.

44.—A plant growing in a window bends towards the light. What inference would you draw as to the effect of light upon the rate of growth?

45.—Procure and examine the structure of the little bladders found on the immersed leaves of the common Bladderwort. Note the action of the trap-door leading into the bladder. Examine also the contents, and make notes of your observations.

46.—Examine the structure and contents of the leaves of the Pitcher-plant. Make drawings and notes.

ORDERS PRESCRIBED FOR STUDY

FOR THE
PRIMARY EXAMINATION.

1. RANUNCULACEÆ.
2. CRUCIFERÆ.
3. MALVACEÆ.
4. LEGUMINOSÆ.
5. ROSACEÆ.
6. SAPINDACEÆ.
7. UMBELLIFERÆ.
8. COMPOSITÆ.
9. LABIATÆ.
10. CUPULIFERÆ.
11. ARACEÆ.
12. LILIACEÆ.
13. IRIDACEÆ.
14. GRAMINEÆ.
15. CONIFERÆ.

OUTLINE OF CLASSIFICATION.

Note. — It will be observed that the arrangement of the groups in this outline is slightly different from that in the key. A comparison of the two are inconsistent will be useful. The presentation is that which is now generally preferred

PLANTS

SERIES............	PHANEROGAMS Flowering	CRYPTOGAMS Flowerless
CLASS	ANGIOSPERMS Seeds enclosed	GYMNOSPERMS Seeds exposed
SUB-CLASS	MONOCOTYLEDONS / DICOTYLEDONS	

DIVISION .... Polypetalous Gamopetalous Apetalous spadiceous Petaloideous Glumaceous Orders Genera Species

Orders Orders Orders Orders Orders Orders
Genera Genera Genera Genera Genera Genera
Species Species Species Species Species Species

PTERIDOPHYTES   BRYOPHYTES   THALLOPHYTES

Ferns Horsetails Club-Mosses   Mosses Liverworts   Fungi Algae

GLOSSARY.

GLOSSARY OF
BOTANICAL TERMS
USED IN PLANT DESCRIPTION.

THE ROOT.

Origin.

PRIMARY: when originating directly from the lower end of the radicle of the embryo (Fig. 1). Such a root is usually (but not always) single, and may send out lateral fibres as it grows; such fibres or branches are included in the primary root.

Annuals and biennials, and many trees, have, as a rule, only primary roots.

SECONDARY: when originating from any other part of the plant than the end of the radicle, as from the sides of stems (Fig. 2), from tubers, rootstocks, bulbs, cuttings, etc.

Perennial herbs, creeping plants, and most shrubs, produce such roots abundantly.

Form.

TAP: having a main central axis, distinctly larger than any of the branches (Fig. 2).

FIBROUS: made up of many similar parts without a distinct central axis (Fig. 4).

A tap root is
  (a) *Conical*, when it gradually tapers from a broad top (Fig. 5).
  (b) *Spindle-shaped* or *fusiform*, when thickest in the middle (Fig. 6).
  (c) *Turnip-shaped* or *napiform*, when nearly globular with an abruptly tapering base (Fig. 7).

Fibrous roots are
  (a) *Of coarse threads*, as in Buttercup.
  (b) *Of fine threads*, as in any common grass.
  (c) *Fascicled* or *clustered* or *tuberous*, when each of the fibres becomes a fleshy mass, as in Peony (Fig. 8).

(In description the Variety may follow the Form on the same line; for example FORM: *Tap, conical.*)

Colour.

In many plants the colour of the root is characteristic, and should always be given in the description.

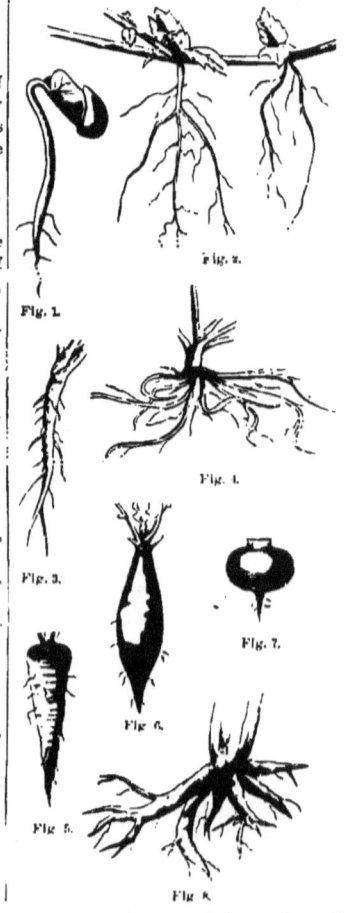

Fig. 1.
Fig. 2.
Fig. 3.
Fig. 4.
Fig. 5.
Fig. 6.
Fig. 7.
Fig. 8.

Position.

SUBTERRANEAN: when, as in most cases, the root is underground.
AËRIAL: when the roots spring from the sides of the stem above ground, as in Poison Ivy, which uses roots for climbing; and in Indian Corn.
AQUATIC: when suspended in water, as in Duckweed.

Duration.

ANNUAL: lasting one season only.
BIENNIAL: lasting two seasons.
PERENNIAL: lasting year after year.

THE STEM.

Class.

EXOGENOUS (or DICOTYLEDONOUS): with the wood in annual layers or rings (Fig. 9).
  Note that plants with exogenous stems have also the following characters:
   (a) The embryo of the seed has more than one (usually two) cotyledons.
   (b) The leaves are net-veined.
   (c) The parts of the flower are usually *not* in threes or sixes, but commonly in fours or fives.
   (d) They have a true bark.
ENDOGENOUS (or MONOCOTYLEDONOUS): with the wood not in rings but scattered through the stem (Fig. 10).
  Plants with endogenous stems have also the following characters:
   (a) The embryo has but one cotyledon.
   (b) The leaves are nearly always straight-veined.
   (c) The parts of the flower are never in fives, but almost invariably in threes or sixes.
   (d) They have no true bark.

Attitude.

ERECT: growing directly upwards.
DECLINED: bending over towards the ground.
PROSTRATE, or PROCUMBENT, or TRAILING: lying flat along the ground.
CREEPING: lying flat, and striking root at intervals (Fig. 11).
DIFFUSE: spreading in all directions.
ASCENDING: growing upwards in a slanting direction.
CLIMBING: when the stem raises itself by means of tendrils (Fig. 12) or leaf-stalks, or hooked prickles, which lay hold of neighbouring plants or other objects.
TWINING: when the stem itself coils round the support (Fig. 13).

Fig. 9.   Fig. 10.

Fig. 11.

Fig. 12.

Fig. 13.

Texture.

Herbaceous: with little or no wood, and dying down to the ground each year.
Woody: as in shrubs and trees.
Suffruticose: woody at the base, but herbaceous at the top.

Position.

Aerial: growing above ground.
Subterranean: growing under ground.
  Of subterranean stems there are the following varieties:
    (a) *Rhizome*, or *Rootstock*: a horizontal, more or less fleshy, perennial underground stem, which produces each season a new bud at its extremity, from which the annual overground stem is developed, as in Trillium, Bloodroot, and most of our early flowering herbs (Fig. 14).

Fig. 14.

    (b) *Tuber*: the thickened end of a rhizome, as the Potato and Artichoke (Fig. 15).
    (c) *Bulb*: a globular mass, usually made up of fleshy scales attached to a short flat stem, as the Lily (Fig. 16) and Onion.

Fig. 15.

    (d) *Corm*: a bulb having the stem part very large compared with the bud or leaf part, as in Indian-Turnip (Fig. 17).
  A plant is described as *acaulescent*, or stemless, when the stem is very short and the leaves spring in a cluster from the surface of the ground, as in Dandelion and Hepatica.

Fig. 16. Fig. 17.

Shape.

Terete: cylindrical (Fig. 18).
Compressed: somewhat flattened (Fig. 19).
Two-edged: Fig. 20.
Square: Fig. 21.
Grooved: Fig. 22.
Winged: Fig. 23.
Striate: with fine lines running lengthwise.

Figs. 18. 19. 20. 21.

Juice.

In some cases the colour or taste of the juice is characteristic, and often of much value. Bloodroot has a red juice, Milkweed and the greater Celandine a yellow juice, Buttercup a colourless bitter juice. Sorrel a sour, etc.

Figs. 22. 23.

Branching.

The stem is
Simple: when it bears no branches, as in Mullein.
Branched: when it bears branches or is divided through to the top, as in most trees.
Decompound: when the branches are several times subdivided.

BOTANICAL TERMS.

WITH RUNNERS: when there are slender branches from the base of the stem which take root at the end, as in Strawberry, etc. (Fig. 11).

WITH STOLONS: when branches bend over so as to reach the ground and take root (Fig. 11).

WITH SUCKERS: when an underground branch sends up a stem at a distance from the parent plant, as in Mint, etc. (Fig. 11).

TENDRILS are sometimes branch-forms, as those of the Grape (Fig. 12).

SPINES, as in Hawthorn, are also branch-forms, stunted and pointed (Fig. 24).

THE LEAF.
Parts.
BLADE: the broad part.

PETIOLE: the leaf-stalk.

STIPULES: two small usually leaf-like pieces, one on each side of the petiole where it joins the stem of the plant (Fig. 25); but sometimes the stipules are in the form of spines, as in Locust, and sometimes they form a tube around the stem, as in Smartweed (Fig. 27).

SHEATH: the tubular petiole which surrounds the stem in many Endogens (Fig. 26).

LIGULE: the thin semi-transparent appendage growing at the top of the sheath in most grasses. It appears to be an upward extension of the lining of the sheath (Fig. 26).

Position.
RADICAL: when arising from the stem at or below the surface of the ground.

CAULINE: all the leaves higher up the stem.

In plants like Dandelion and Hepatica *all* the leaves are radical. In Buttercup and Shepherd's Purse there are both kinds (Fig. 28).

Arrangement.
ALTERNATE: when only one leaf springs from a node, or joint of the stem (Fig. 29).

OPPOSITE: when two leaves spring from each node on opposite sides of the stem; and opposite leaves are *decussate* when each pair is at right angles to the next pair (Fig. 30).

WHORLED, or VERTICILLATE: when three or more leaves spring from a node (Fig. 31).

FASCICULATE: when there are several leaves in a bundle, as in Pine, Larch, etc. (Fig. 32).

Division.

Simple: when the blade is in one piece, however deeply it may be cut.

Compound: when the blade is in two or more distinct pieces, which are then known as *leaflets*.

A compound leaf is

(a) *Pinnate*: when the leaflets are arranged on each side of a central or mid rib; and such a pinnate leaf will be *odd-pinnate* if there is an odd leaflet at the end (Fig. 33); *abruptly-pinnate* if there is not a terminal leaflet (Fig. 34); and *pinnate with a tendril* if the mid-rib ends in a tendril, as in Pea, etc. (Fig. 35).

Again: the leaf is *twice-pinnate* if the primary divisions are themselves pinnate (Fig. 36); *thrice-pinnate* if the subdivision is carried through another stage; and *decompound* if still more divided.

It is *interruptedly-pinnate* if, as in Tomato, there are small leaflets interspersed among the larger ones (Fig. 37).

(b) *Palmate*: if the leaflets are spread out from the end of the petiole, like fingers (Fig. 38).

A compound leaf is further described by mentioning the number and form of the leaflets. (An example of the complete description of a compound leaf is given at the beginning of the leaf-schedules later on).

Venation.

Straight-veined: when the veins run nearly parallel, either from end to end of the leaf, as in grasses (Fig. 39), or from a central rib to the margin, as in Calla (Fig. 40).

Net-veined: when the veins run in all directions, forming a network. Such a leaf is

(a) *pinnately net-veined*: when there a distinct central rib with the smaller veins branching from it on each side, Fig. 41; and

(b) *palmately net-veined*: when there are several chief ribs radiating from the end of the petiole. Fig. 42.

Outline.

1 Of leaves nearly alike at both ends.

Filiform: thread-like, as in Asparagus.

Acicular: needle-shaped, as in Pine (Fig. 43).

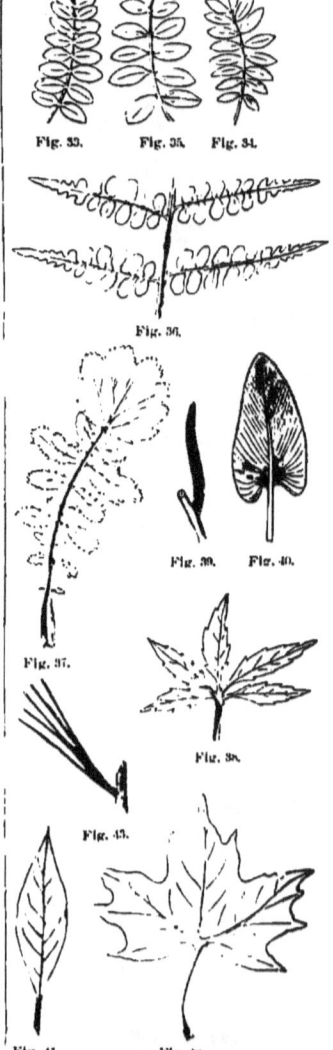

Fig. 33. Fig. 35. Fig. 34.

Fig. 36.

Fig. 39. Fig. 40.

Fig. 37.

Fig. 38.

Fig. 43.

Fig. 41. Fig. 42.

BOTANICAL TERMS.

LINEAR: narrow compared with the length (Fig. 44).
OBLONG: not more than three times as long as broad, and with sides inclined to be straight (Fig. 45).
OVAL, or ELLIPTICAL: not more than twice as long as broad (Fig. 46).
ORBICULAR: round, or nearly so (Fig. 47).

2. Of leaves broadest below the middle.

SUBULATE: awl-shaped (Fig. 48).
LANCEOLATE: as in Fig. 49.
OVATE: as in Fig. 50.
DELTOID: about as broad as long, and rather triangular (Fig. 51).

3. Of leaves broadest above the middle.

OBLANCEOLATE: the reverse of lanceolate (Fig. 52).
SPATHULATE: like the last, but more rounded at the top (Fig. 53).
OBOVATE: the reverse of ovate (Fig. 54).
WEDGE-SHAPED, or CUNEATE: like the last, but with the end more flattened and the margins nearly straight (Fig. 55).

In describing outlines, it will often be necessary to combine terms, as for example: *linear-oblong, linear-lanceolate, oblong-ovate*, etc., as the case may require.

Margin.

ENTIRE: not indented in any way (Fig. 56).
SERRATE: with sharp teeth pointing forward like the teeth of a saw (Fig. 57).
SERRULATE: very finely serrate (Fig. 58).
DENTATE: with teeth pointing outward (Fig. 59).
CRENATE: with teeth rounded at the point (Fig. 60).

A margin may also be *doubly-serrate* (Fig. 61), *doubly-dentate*, or *doubly-crenate* (Fig. 62), when the larger teeth are themselves serrate, or dentate, or crenate.

SINUATE: deeply wavy (Fig. 63).
CILIATE: with a fringe of hairs.
REVOLUTE: with the edge turned back.
REPAND: like the edge of an expanded umbrella (Fig. 64).
PINNATIFID: when the edge of a pinnately-veined leaf is very deeply lobed (Fig. 65).
BI-PINNATIFID: when the first lobes are themselves pinnatifid (Fig. 66).

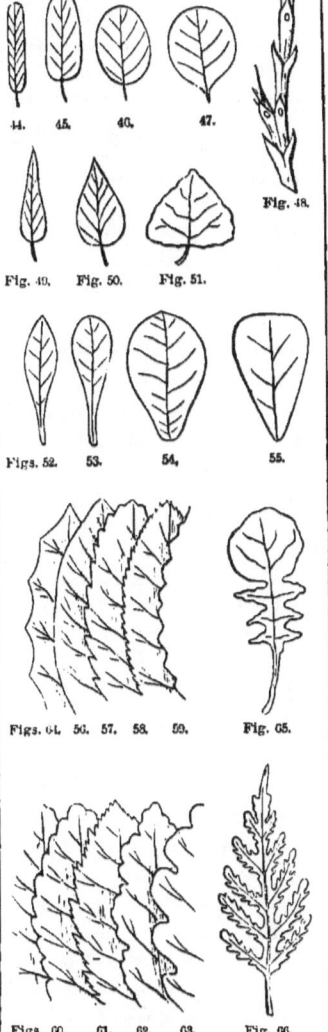

GLOSSARY OF

PALMATIFID: when the edge of a palmately-veined leaf is very deeply lobed (Fig. 67).

PECTINATE: when the edge somewhat resembles the teeth of a comb.

LYRATE: pinnatifid, with a very large lobe at the end (Fig. 65).

RUNCINATE: pinnatifid, with the lobes pointing backwards, as in Dandelion (Fig. 68).

PEDATE: palmatifid, with the lobes at the base two-cleft (Fig. 69).

MULTIFID: cut into many fine segments or lobes, as in Milfoil.

Apex.

ACUMINATE: running out to a long slender point (Fig. 70).

ACUTE: making an acute angle (Fig. 71).

OBTUSE: making an obtuse angle; blunt (Fig. 72).

TRUNCATE: as if the end were cut off square (Fig. 73).

RETUSE: with the end slightly indented (Fig. 74).

EMARGINATE: with a distinct notch (Fig. 75).

OBCORDATE: rather deeply notched (Fig. 76).

CUSPIDATE: with a short but distinctly tapering point (Fig. 77).

MUCRONATE: with a fine sharp point projecting beyond the end of the mid-rib (Fig. 78).

ARISTATE: tipped with a bristle.

Base.

ACUTE: making an acute angle (Fig. 79).

OBTUSE: making an obtuse angle; blunt (Fig. 46).

TAPERING: with a long and slender base (Fig. 80).

CORDATE: rounded and notched (Fig. 67).

AURICULATE: with two small rounded lobes (Fig. 81).

SAGITTATE: with sharp lobes pointing downwards (Fig. 83).

HASTATE: with sharp lobes pointing outwards (Fig. 84).

PELTATE: when the petiole is attached, not to the edge, but to the under surface (Fig. 85).

RENIFORM: with very large rounded lobes (Fig. 86).

CLASPING: when the leaf is sessile, and the lobes are close against the stem on each side (Fig. 82).

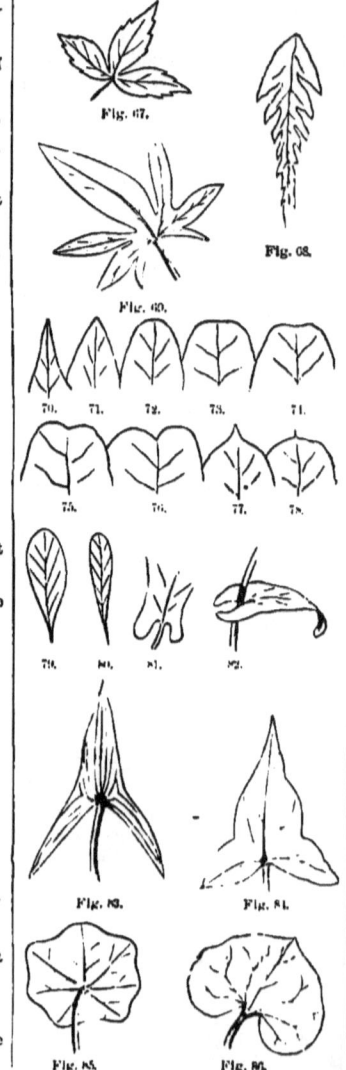

PERFOLIATE: when the lobes of a sessile leaf grow together at the back of the stem, so that the stem seems to pass through the leaf (Fig. 87).
CONNATE, or CONNATE-PERFOLIATE: when two opposite sessile leaves grow together by their bases (Fig. 88).
DECURRENT: when the lobes of a sessile leaf grow down the sides of the stem (Fig. 89).

Surface.

(The student should use his lens in determining the character of the surface of either stem or leaf.)
SMOOTH, or GLABROUS: entirely without hairs.
GLAUCOUS: covered with a bloom which may be rubbed off with the fingers, as in Cabbage.
PUNCTATE: showing transparent dots when held up to the light, as in St. John's Wort.
SCABROUS: rough, but without hairs.
PUBESCENT: covered with fine soft short hairs.
VILLOUS: with long soft hairs.
TOMENTOSE: with matted hairs.
SERICEOUS: with silky hairs.
HOARY: with white down.
HISPID: with stiff hairs.
SPINOUS: with scattered spines.
RUGOSE: wrinkled.
CILIATE: with hairs on the *edge*.

Fig. 87.

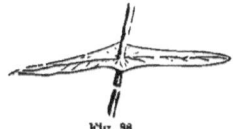

Fig. 88.

Colour.

The colour of the leaf must be described by an appropriate term, and if, as is often the case, the two surfaces differ in colour, this fact must be noted.

Texture.

Leaves differ very much in texture. Some are very thin and soft, others almost leathery, while others again are very thick and fleshy. In describing a leaf, judgment must be exercised in selecting a suitable term.

Fig. 89.

Duration

FUGACIOUS, or CADUCOUS: falling off early in summer.
DECIDUOUS: falling off in autumn, as in most trees and shrubs.
PERSISTENT, or EVERGREEN: remaining at least a year on the plant.

Vernation, or mode of folding in the bud.

CONDUPLICATE: doubled lengthwise. Shown in cross-section in Fig. 90.
PLICATE: folded like a fan, as in Mallow (Fig. 91).
CONVOLUTE: rolled from one edge to the other (Fig. 92).

Fig. 90.   Fig. 91.   Fig. 92.

INVOLUTE: rolled inward from both edges (Fig. 93).
REVOLUTE: rolled backward from both edges (Fig. 94).
CIRCINATE: coiled from the apex, as in Ferns (Fig. 95).
EQUITANT: each leaf doubled lengthwise and astride of the next leaf within, as in Iris (Fig. 96).

Fig. 93. Fig. 94. Fig. 95.

INFLORESCENCE.

Arrangement of the Flowers or Flower-clusters on the stem.

Mode.

TERMINAL: when the separate flowers are on the ends of stems or branches.

Terminal Inflorescence is also known as DETERMINATE, or DEFINITE, or CYMOSE, or CENTRIFUGAL, and it is

(a) *Solitary*: when a single flower terminates the stem, as in Tulip and Hepatica. In other words the flowers do not form a cluster (Fig. 97).

(b) *A Cyme*: when the flowers are in a cluster of which the central flower (on the end of the main stem) is the earliest (Fig. 98), as in Chickweed and Sweet-William. In Chickweed the cyme is *loose*, and in Sweet-William it is *dense*.

(Special cases of Cymes arising from the axils of leaves are referred to below under the head of Mixed Inflorescence.)

AXILLARY: when the separate flowers spring from the axils of leaves or bracts.

Axillary Inflorescence is also known as LATERAL or INDETERMINATE, or INDEFINITE, or RACEMOSE, or BOTRYOSE, or CENTRIPETAL; and it is

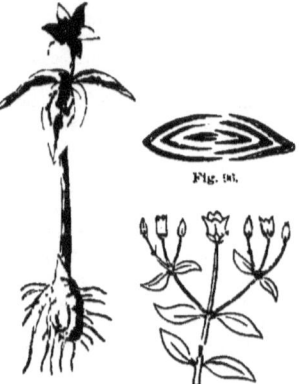

Fig. 96.

Fig. 97. Fig. 98.

(a) *Solitary*: when the flowers are produced singly in the axils of ordinary leaves (not bracts), as in Morning-Glory, etc. The flowers do not form a cluster.

(b) *A Raceme*: when the flowers form a rather long cluster, each flower being in the axil of a bract, and having a pedicel (little stalk) of its own (Fig. 99).

(In plants of the Cross family the bracts are absent.)

(c) *A Spike*: when the separate flowers are sessile, or nearly so, along the main axis, as in Hollyhock, etc. Fig. 100.

Fig. 99. Fig. 102.

(d) *A Head*: when the axis of the cluster is short, and the flowers are closely packed together, as in Clover and Thistle, etc.

(e) *An Umbel*: when the pedicels of the flowers are of the same length and arise from the same point (Fig. 101).

(f) *A Corymb*: when the pedicels arise from different points on the stem, but the flowers reach the same level above (Fig. 102).

Fig. 100. Fig. 101.

The Raceme, Umbel, and Corymb may be compound, as shown in Figs. 103 (compound Raceme) and 104 (compound Umbel).

(g) *A Catkin:* when the flowers (usually imperfect) arise from scale-like bracts along a slender axis. The Catkin is thus a special kind of spike (Fig. 105).

(h) *A Spadix:* when the flowers (often imperfect) are arranged in a spike-like cluster on a fleshy axis, as in Indian-Turnip (Figs. 106 and 107).

The Spadix is usually surrounded by a large showy bract called a spathe (Fig. 108).

MIXED: when axillary and terminal forms are combined. For example, in many Composites the inflorescence is terminal or cymose *as to the heads themselves,* while each head separately is always axillary or lateral as to the development of the florets of which it is made up. The chief varieties of mixed inflorescence are

(a) *The Thyrse:* a cluster like that of Lilac, in which the primary branches are lateral, and the secondary cymose.

(b) *Verticillaster:* a cluster like that of Catnip and Mint flowers generally, where two dense cymes form in the axils of opposite leaves, giving the appearance of a whorl.

In connection with inflorescence the following terms should be noticed:

Peduncle: the flower-stalk, or in the case of clusters the stalk supporting the whole cluster.

Pedicel: the separate stalk of each flower in a cluster.

Scape: a leafless flower-stalk rising from the ground or near it, as in Tulip and Dandelion.

Bract: a foliage-leaf, differing from the ordinary leaves of the plant in size, shape or colour, and found under the flower or flower-cluster.

Bractlet: a secondary bract, as seen on the pedicels in Fig. 103.

Involucre: a circle of bracts, such as the outer leaves of Composite flowers like Dandelion, etc. (Fig. 109).

Involucel: a secondary or minor involucre or circle of bractlets, such as is commonly found under the small clusters of a compound umbel (Fig. 104).

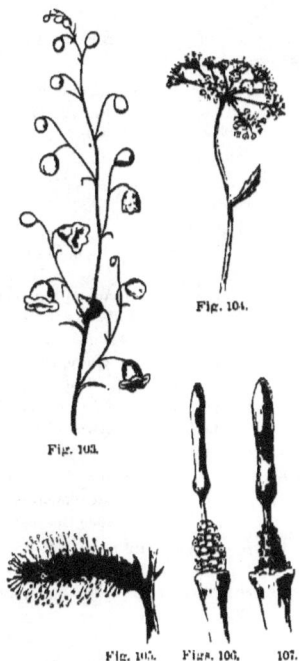

Fig. 104.

Fig. 103.

Fig. 105. Figs. 106. 107.

Fig. 108. Fig. 109.

THE FLOWER.

Parts:

CALYX: the outer set of flower-leaves, usually green or greenish, as in Buttercup (Fig. 110). The pieces of which the calyx is made up are called *sepals*.

COROLLA: the second set of flower-leaves, immediately within the calyx, and usually some other colour than green (Fig. 110). The pieces of which the corolla is made up are called *Petals*.

These two sets taken together are known as the *Floral Envelopes*, and also as the *Perianth*, but the latter term is generally restricted to the flowers of Monocotyledons, such as Lilies, where the parts are very much alike. Some flowers of Dicotyledons, such as Marsh-Marigold (Fig. 111), have only *one* set of floral envelopes, and this is then nearly always the calyx, no matter what its colour is.

STAMENS or ANDRŒCIUM: the third set of flower-leaves, appearing as thread-like stalks with thickened ends (Fig. 112). These produce the *pollen*.

PISTIL or GYNŒCIUM: the central organ of the flower which bears the seed. It may be in several pieces, as in Buttercup (Fig. 113), or in one piece as in Shepherd's Purse (Fig. 114).

These last two sets taken together are the *Essential Organs* of the flower. They alone are directly concerned in the production of seed. The floral envelopes protect the essential organs, and attract insects which help to distribute the pollen.

RECEPTACLE: the enlarged top of the peduncle to which the parts of the flower are attached.

Note also the following points: Flowers are

(a) PERFECT, if they have both stamens and pistil, whether calyx and corolla are present or not.

(b) IMPERFECT, if either stamens or pistil is wanting. And imperfect flowers are

(1) *Staminate*, if they bear stamens, but not pistil, as in Willow (Fig. 115).

(2) *Pistillate*, if they bear pistils, but not stamens, as in Willow (Fig. 116).

(3) *Neutral*, if both stamens and pistil are absent.

4. *Monœcious*, if staminate and pistillate flowers are borne on the same plant, as in Cucumber and Indian Corn.

Fig. 110.

Fig. 111.

Fig. 112.

Fig. 113.   Fig. 114.

Fig. 115.   Fig. 116.

(5) *Diœcious*, when staminate and pistillate flowers are borne on different plants, as in Willow.

(c) POLYGAMOUS, when there is a mixture of perfect and imperfect flowers.

(d) COMPLETE, if all four parts, viz: calyx, corolla, stamens, and pistil, are present.

(e) INCOMPLETE, if any one or more of the four sets are wanting. Incomplete flowers are *achlamydeous* when calyx and corolla are both wanting, as in Willow.

(f) SYMMETRICAL, if the different sets consist of the same number of pieces each, or of a multiple of the same number, for example: 4 sepals, 4 petals, 8 stamens, 4 carpels.

(g) UNSYMMETRICAL, if there are not the same number of pieces (or a multiple of the same number) in each set.

(h) REGULAR: when the pieces of each set are alike in size and shape, as in Buttercup (Fig. 110).

(i) IRREGULAR: when the pieces of each set are not alike in size and shape, as in Sweet Pea, Orchid, etc. (Fig. 117).

Fig. 117.

THE CALYX.

Cohesion (union of like parts).

POLYSEPALOUS: with the sepals entirely distinct from each other, so that they can be pulled off separately, as in Buttercup (Fig. 110).

GAMOSEPALOUS: when the sepals are all united together (Fig. 118).

Fig. 118.

···· Pappus.
Fig. 119.

The following terms are applicable to the gamosepalous calyx:

(a) *The Tube*: the lower united part (Fig. 118).

(b) *The Limb*: the upper separated part (Fig. 118), made up of lobes or teeth. In many composite flowers the limb is pappose, consisting of fine bristles (Fig. 119).

(c) *The Throat*: the entrance to the calyx-tube.

Adhesion (union of unlike parts).

INFERIOR: when the calyx is plainly beneath the ovary and free from it, as in Buttercup, etc. (Fig. 112).

SUPERIOR: when the calyx-tube grows fast to the outside of the ovary and the limb rises above it, as in Apple, etc. (Fig. 120).

Fig. 120.

GLOSSARY OF

Duration.

FUGACIOUS or CADUCOUS: falling off as soon as the flower opens, as in Bloodroot and Poppy.

DECIDUOUS: falling off about the same time as the corolla and stamens, as in Buttercup.

PERSISTENT: remaining after the corolla has fallen off, as in Hollyhock and Sweet-Brier (Fig. 121).

As the sepals are only modified leaf-forms, they may be further described by means of the terms already explained for leaves, such as *lanceolate*, *pubescent*, etc.

THE COROLLA.

Cohesion.

POLYPETALOUS: when the petals are entirely distinct from each other, as in Buttercup (Fig. 110). In such petals two parts may often be distinguished, a broad upper part, the *limb*, and a narrower lower part, the *claw* (Fig. 122).

GAMOPETALOUS: when the petals are grown together in however slight a degree, so that the corolla may be pulled off in one piece, as in Convolvulus, etc. (Fig. 123).

The terms *tube*, *limb* and *throat* are applicable to such corollas, as well as to the gamosepalous calyx.

A gamopetalous corolla is further described by stating its *Form*. It is

a) *Tubular*, when of nearly the same width from top to bottom (Fig. 124).

b) *Funnel-shaped*, when the tube spreads out gradually into a wide border (Fig. 123).

c) *Campanulate*, or bell-shaped, when the tube is short and wide, with a slightly spreading border (Fig. 125).

d) *Salver-shaped*, when the tube is long and narrow with a spreading border at right angles to it (Fig. 126).

e) *Rotate*, when the tube is very short with a spreading border (Fig. 127).

f) *Urceolate*, or urn-shaped, when the tube is swollen below and contracted at the mouth (Fig. 128).

g) *Labiate*, when distinctly two-lipped as in Catnip and Turtle-head (Fig. 129).

A Labiate corolla is further described as

1. *Ringent*, when the mouth is wide open (Fig. 129).

2. *Personate*, when the mouth is closed by an upward projection of the lower lip called the *palate* (Fig. 130).

3. *Ligulate*, when one side of the tube is prolonged into a ribbon or strap, as in Dandelion (Fig. 119).

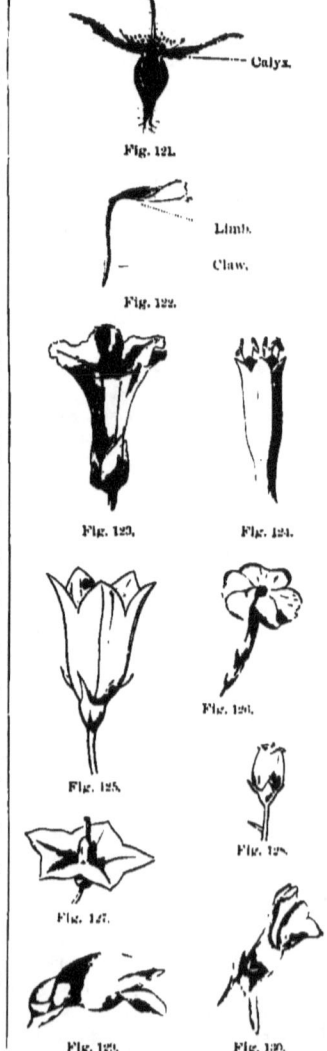

The form of corolla peculiar to plants of the Pulse Family is known as

PAPILIONACEOUS (Fig. 131): it consists of five petals; an upper large one (the *standard*), two side ones (the *wings*), and two lower ones which are united together to form the *keel*.

Finally, both gamopetalous and polypetalous corollas may have one or more petals prolonged into *spurs* at the base (Fig. 130).

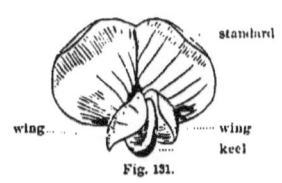

Fig. 131.

Adhesion.

HYPOGYNOUS: when inserted on the receptacle, under the ovary and free from it (Fig. 132, c).

PERIGYNOUS: when inserted on the calyx (Fig. 133, c.)

EPIGYNOUS: when inserted on the top of the ovary (Fig. 134, c).

The corolla should be further described by giving the shape, colour and size of the petals, using the ordinary terms.

In describing the flowers of monocotyledons having a *coloured perianth*, use the following terms for cohesion:

POLYPHYLLOUS: when the pieces of the perianth are entirely separate.

GAMOPHYLLOUS: when the pieces of the perianth are united.

For adhesion, use the terms *superior* and *inferior*, as explained above for the calyx.

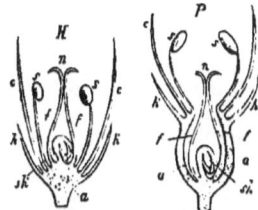

Fig. 132.   Fig. 133.

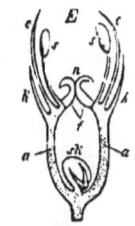

Fig. 134.

Æstivation.

This term is applicable to both calyx and corolla, and means the mode in which these organs are folded in the bud. It is

(a) *Valvate*, if the edges of the parts meet without overlapping (Fig. 135), as in the calyx of Mallow.

(b) *Convolute*, if the members of a set overlap so that each has one edge covered and the other uncovered (Fig. 136), as in the corolla of Mallow.

(c) *Imbricate*, when the members of a set overlap so that at least one piece has both edges uncovered and at least one piece has both edges covered (Fig. 137), as in Apple.

(d) *Plicate* or *plaited*, applied to the folding of gamopetalous corollas. The plaits may overlap in the convolute manner, as in Fig. 138; they are then said to be *supervolute*.

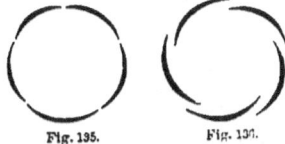

Fig. 135.   Fig. 136.

Fig. 137.   Fig. 138.

THE STAMENS or ANDRŒCIUM.

Parts.

FILAMENT: the lower stalk-like part; it supports the anther (Fig. 140). Stamens are

(a) *Exserted*, if the filaments are so long that the anthers protrude beyond the perianth (Fig. 141).

(b) *Included*, if the filaments are not long enough to raise the anthers beyond the perianth (Fig. 126).

(c) *Sessile*, if the filaments are absent (Fig. 142).

ANTHER: the swollen upper part, consisting of one or more (usually two) sacs or cells which contain the pollen (Fig. 139).

One surface of the anther is usually more deeply grooved than the other; this is the *face*, the other being the *back*.

An anther is

(a) *Introrse*, if the face is toward the centre of the flower.

(b) *Extrorse*, if turned outwards.

Attachment of the Anther.

The anther may be attached to the filament in three ways. It is

(a) *Innate*, if its lower end rests on the top of the filament (Fig. 143).

(b) *Adnate*, if the back of the anther lies with its whole length against and attached to the filament (Fig. 144).

(c) *Versatile*, if the end of the filament is attached to a point on the back of the anther, so that the latter swings easily (Fig. 145).

Dehiscence of the Anther.

The anther may open in several ways to allow the escape of the pollen. The dehiscence is

(a) *Longitudinal*, when the anther-cell opens from top to bottom, a lateral slit (Fig. 146). This is the usual mode.

(b) *By valves*, when the side of the anther-cell turns up like a hinge (Fig. 147).

(c) *By pores*, when the pollen escapes through a minute opening at the top of the anther-cell (Fig. 148).

CONNECTIVE: the second part between the anther-cells. Occasionally the connective is reduced or wanting.

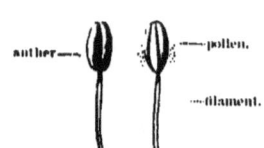

Fig. 139. Fig. 140.

Fig. 141. Fig. 142.

Fig. 143. Fig. 144.

Fig. 145.

Figs. 146, 147, 148.

BOTANICAL TERMS.

POLLEN: the minute grains (alike in the same plant, but very different in different plants) contained in the anther-cells, commonly resembling a loose dust or powder (Fig. 140), but sometimes cohering in sticky masses *(pollinia)*, as in Orchis (Fig. 149).

Pollen-grains are plant-cells having two coats, and enclosing a thickish liquid. Fig. 150 shows a single pollen-grain with its inner coat growing out in the form of a tube.

The pollen is the *essential* part of the stamen. The pupil should examine with a good microscope various kinds of pollen-grains, and make drawings of them.

Number.

If the stamens are not more than *ten* in number, the exact number should be stated. If more than ten, they are *numerous* or *indefinite*, and this is indicated by the sign ∞ in the proper column of the descriptive table.

Cohesion.

If the stamens are entirely separate from each other, their cohesion (or the absence of it) is described by prefixing to the ending *-androus* the Greek prefix corresponding to the number of stamens present, as follows:

1.	2.	3.	4.	5.	6.	7.
mon-	di-	tri-	tetr-	pent-	hex-	hept-
8.	9.	10.		more than 10.		
oct-	enne-	dec-		poly- androus.		

The cohesion is

DIDYNAMOUS: if there are four stamens, two long and two short (Fig. 151).

TETRADYNAMOUS: if there are six stamens, four long and two short (Fig. 152).

MONADELPHOUS: when all the filaments are grown together, leaving the anthers separate, as in Mallow (Fig. 153).

DIADELPHOUS: when the filaments are grown together in two sets, as in Pea (Fig. 154).

TRIADELPHOUS: when the filaments are grown together in three sets, as in St. John's Wort (Fig. 155).

POLYADELPHOUS: when the filaments are grown together in more than three sets.

SYNGENESIOUS: when all the *anthers* are grown together, leaving the filaments separate, as in Dandelion (Fig. 155).

Adhesion.

HYPOGYNOUS: when inserted on the receptacle under the ovary (Fig. 132, *s*).

PERIGYNOUS: when inserted on the calyx (Fig. 133, *s*).

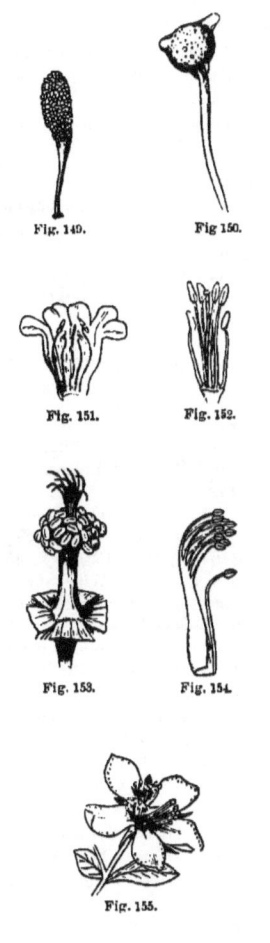

Fig. 149. Fig 150.
Fig. 151. Fig. 152.
Fig. 153. Fig. 154.
Fig. 155.
Fig. 156.

EPIGYNOUS: when inserted on the ovary (Fig. 134, s).
EPIPETALOUS: when inserted on the corolla (Fig. 151).
EPIPHYLLOUS: when inserted on the perianth (in Monocotyledons).
GYNANDROUS: when inserted on the style, as in Orchids (Fig. 157).

Situation.

It is important to note the position of the stamens with reference to the petals when they are of the same number as the latter. They may be
(a) *Alternate* with the petals.
(b) *Opposite* the petals.

THE PISTIL OR GYNŒCIUM.

Parts.

CARPELS: the pieces, either distinct or combined together, which make up the whole pistil. The pistil is
(a) *Simple*, if it consists of one carpel only, as in Pea (Fig. 158).
(b) *Compound*, if it consists of two or more carpels, either separate from each other *(apocarpous)* as in Buttercup (Fig. 159), or combined together *(syncarpous)* as in Fig. 160. When several carpels are combined, the number is very commonly indicated by seams or *sutures* on the outside of the ovary.

Whether composed of one carpel or several combined, the pistil may have the following parts:

OVARY: the lower swollen part, containing the ovule or ovules which develope into seeds (Fig. 160). The ovary may be one-celled even when compound (Fig. 161), or several-celled (Fig. 160). In the latter case the separating walls are called *dissepiments*, and the cells are often spoken of as *loculi* (sing. *loculus*).

STYLE: the narrow part above the ovary (Fig. 160). A compound pistil may have several styles, as in Fig. 162.

STIGMA: the moist roughish upper end of the style. This part differs from the rest of the pistil in having no skin or epidermis (Fig. 163).

The stigma is
(a) *Capitate*, if it forms a knob or button on the end of the style Fig. 164.
(b) *Plumose*, if of a feathery appearance as in grasses (Fig. 165).
(c) *Petaloid*, if leaf like and coloured, as in Iris (Fig. 166).

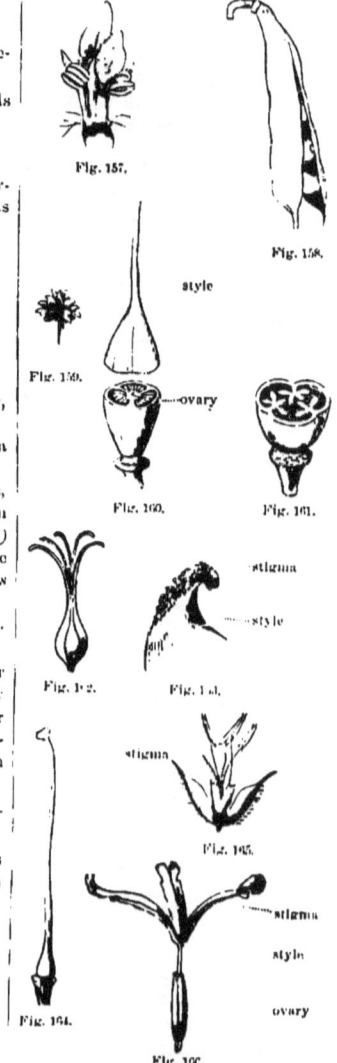

Note that the essential parts of the pistil are the ovary or seed-bearing part, and the stigma which receives the pollen.

The style is often wanting, and then the stigma is *sessile*.

An exceptional pistil is found in *gymnospermous* plants like the Pine. Here the ovules are not enclosed, but are attached to the inner face of an open leaf or scale, the scales forming a *cone* (Figs. 167, 168, 169).

Fig. 167.

Cohesion.

APOCARPOUS: when the carpels are not united together in any way (Fig. 159).

SYNCARPOUS: when the carpels are grown together in any degree (Fig. 160). They may be united merely at the base of the ovary, or to the top of the style.

Figs. 168, 169.

Adhesion.

SUPERIOR: when entirely free from the calyx (Fig. 132, *f*), as in Buttercup, Shepherd's Purse, etc.

INFERIOR: when surrounded by the calyx-tube which grows fast to it (Fig. 134, *f*), as in Apple and Fuchsia.

THE OVULE.

Definition.

Ovules are the bodies which, after fertilization by the pollen, develope into seeds.

Placentation.

By this term is meant the arrangement of the placentas, or projections in the interior of the ovary upon which the ovules grow. Placentation is

(a) *Marginal*, in a simple pistil like that of Pea, the placenta being on one seam or *suture* (Fig. 158).

(b) *Axile* or *Central*, when the pistil is compound, and the dissepiments meet in the centre of the ovary (Fig. 160.)

(c) *Parietal*, when the compound ovary is one-celled and the ovules are borne on the walls (Fig. 161).

(d) *Free Central*, when the ovary is one-celled, and the ovules are borne on a column which rises from the bottom of the cell (Figs. 170, 171).

Figs. 170, 171.

Parts of the Ovule.

FUNICULUS: the stalk by which the ovule is attached to the placenta (Fig. 173, *f*). If this stalk is absent the ovule is *sessile*.

PRIMINE: the outer coat of the ovule (Fig. 172, *ai*).

SECUNDINE: the inner coat (Fig. 172, *ii*).

MICROPYLE: the minute opening through the two coats (Fig. 172, *m*).

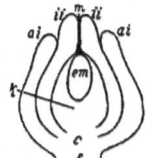

Fig. 172.

NUCLEUS: the body of the ovule within the coats (Fig. 172, k).
EMBRYO-SAC: the large cell in the nucleus in which the young plant is developed (Fig. 172, em).
CHALAZA: the portion where the two coats are blended together (Fig. 172, c).

Kinds of Ovule.

ORTHOTROPOUS: when the ovule is erect, and the micropyle is as far as possible from the funiculus or point of attachment (Fig. 172).

ANATROPOUS: when the ovule is completely inverted or bent upon itself so as to bring the micropyle close to the point of attachment (Fig. 173). In this case the funiculus becomes fused with the primine on one side, forming the raphe (Fig. 173, r).

CAMPYLOTROPOUS: when the ovule is half bent over (Fig. 174).

Fertilization.

Ovules are converted into seeds by the action of pollen upon them. Pollen grains fall upon the stigma which is moist and retains them. The grains begin to grow as shown in Fig. 150, the inner coat being protruded as a slender tube which makes its way down through the style into the ovary, and then through the micropyle of the ovule, finally attaching itself to the surface of the embryo-sac, and carrying the contents of the pollen-grain with it. Presently growth begins inside the embryo-sac, and soon the embryo is formed. It is the presence of the embryo which marks the distinction between an ovule and a seed.

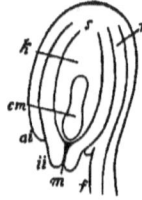
Fig. 173.

In most cases the ovule is fertilized by pollen brought from another flower of the same species (cross-fertilization), because very commonly the pollen of its own flower is ready either too soon or too late to be of use; that is, the pollen and the stigma in the same flower do not commonly mature at the same time. Plants are

Entomophilous, when they depend upon insects to carry the pollen from flower to flower, and

Anemophilous, when this service is performed by the wind.

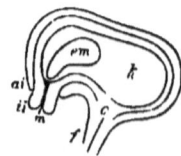
Fig. 174.

THE FRUIT.

Definition.

The fruit is the ripened pistil together with any other part, such as the calyx or receptacle, which may be adherent to it. If there are no such adherent parts the fruit is a *true fruit*, consisting wholly of the ripened ovary with the seeds; otherwise it is a *pseudocarp* or *spurious fruit*, as in Apple, Strawberry and Rose.

The essential parts of the fruit are
(a) *The Seed*, or matured ovule, and
(b) *The Pericarp*, or matured ovary, within which the seeds are contained.

The Pericarp is in three layers:
(a) *The Epicarp* (or Exocarp), the outer layer.
(b) *The Mesocarp* (or Sarcocarp), the middle layer.
(c) *The Endocarp*, the inner layer.

Kinds of Fruit.

A.—DRY FRUITS: those whose pericarp remains thin, and becomes dry and hard at maturity. Such fruits are
(1) *Dehiscent*, when the pericarp opens so as to allow the seeds to escape.
(2) *Indehiscent*, when the pericarp does not so open.

Dry Dehiscent Fruits.

(a) *Follicle*, a fruit of a single carpel, which opens down one edge only, as in Marsh-Marigold and Peony (Fig. 175).

(b) *Legume*, a fruit of a single carpel, which opens down both edges (dorsal or outer and ventral or inner sutures), as in Pea and Bean (Fig. 176).

The *Loment* is a special form of legume. It is made up of a number of one-seeded joints which separate from each other when ripe; each joint, as a rule, remaining closed (Fig. 177).

(c) *Silique*, a syncarpous fruit of two carpels divided by a thin partition, from which the carpels fall away when ripe, leaving the placentas and seeds around the edge of the partition (Fig. 178).

(d) *Silicle*, a fruit of the same construction as the silique, and differing only in shape; the silique being considerably longer than broad, as in Stock (Fig. 178), and the silicle being nearly or quite as broad as long, as in Shepherd's Purse (Figs. 179, 180).

(e) *Pyxis*, a fruit which opens by a horizontal seam, so that the top comes off like a lid, as in Purslane (Fig. 181).

(f) *Capsule*, a syncarpous fruit which normally splits at maturity, either wholly or partially, into as many pieces as there are carpels.

The Dehiscence of the Capsule is
SEPTICIDAL: when the splitting takes place in the line of the dissepiments (Fig. 182).

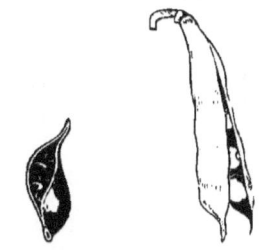

Fig. 175.   Fig. 176.

Fig. 177.   Fig. 178.

Fig. 179.   Fig. 180.   Fig. 181.

Fig. 182.

LOCULICIDAL: when the splitting takes place in the middle of the wall of each carpel, that is, along the dorsal sutures (Fig. 183).

SEPTIFRAGAL: when the walls split away from the partition, leaving the latter standing (Fig. 184).

CIRCUMCISSILE: when the top of the pericarp comes off like a lid (Fig. 181).

BY PORES: when the seeds escape through small openings near the top of the capsule, as in Poppy.

Dry Indehiscent Fruits.

(a) *Achene*, a dry indehiscent one-seeded fruit, having the pericarp free from the seed, as in Buttercup (Figs. 185, 186), and all Composites.

(b) *Caryopsis* or *Grain*, a dry indehiscent one-seeded fruit, having the pericarp adherent to the seed, as in the Oat (Fig. 187), and Grasses generally.

(c) *Nut*, a dry indehiscent one-seeded fruit with a hard thick pericarp, and usually the product of a syncarpous pistil, in which all the cells and seeds but one have disappeared during growth.

The nut is often accompanied by a *Cupule* or hardened involucre, as in the Acorn (Fig. 188), Beech-nut and Hazel-nut.

(d) *Utricle*, like an Achene, but with a very thin loose pericarp (Fig. 189).

(e) *Schizocarp*, a dry indehiscent two-several-seeded fruit, which breaks up at maturity into one-seeded pieces (carpels), each of which, however, remains closed, as in Mallow (Fig. 190), and all Umbelliferous plants (Fig. 191).

(f) *Samara* or *Key*, a dry indehiscent one-seeded fruit, with a flat wing, as in Elm (Fig. 192), and Ash. The Maple (Fig. 193) has a double samara, which splits into two pieces at maturity, and so is a true schizocarp.

B. – FLESHY FRUITS (all indehiscent):

Drupe or *Stone-fruit*, a fleshy fruit, having a very hard endocarp (the *putamen*), which encloses the seed (1 2 to one only), a thick and usually juicy mesocarp, and a thin outer skin or epicarp, as the Plum, Cherry, Walnut and Peach (Fig. 194).

b. *Berry*, a fleshy fruit, having a soft and juicy endocarp, in which the seeds are embedded, as the Grape, Tomato, Currant, etc. (Fig. 195).

The Orange is a special kind of berry known as a *Hesperidium*.

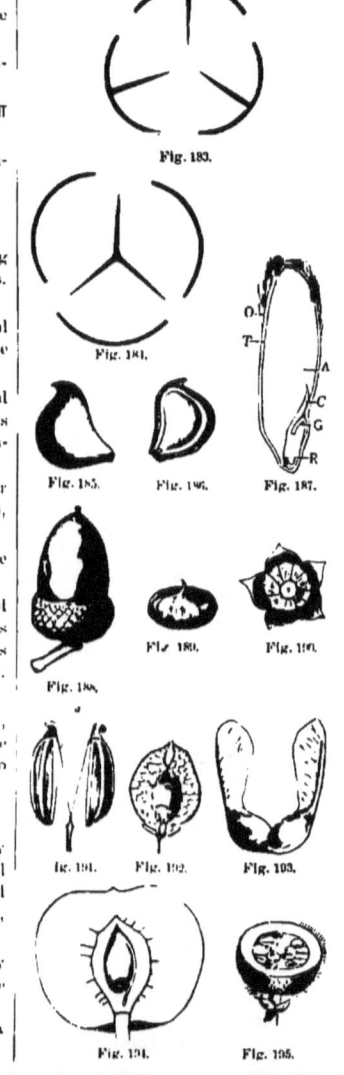

Fig. 183.
Fig. 184.
Fig. 185. Fig. 186. Fig. 187.
Fig. 189. Fig. 190.
Fig. 188.
Fig. 191. Fig. 192. Fig. 193.
Fig. 194. Fig. 195.

(c) *Gourd* or *Pepo*, a modified berry, having a hard rind, as in Pumpkin, Squash, etc.

(d) *Pome*, a fleshy pseudocarp, the product of a syncarpous pistil, in which the fleshy layer consists chiefly of an enlarged calyx-tube, as in Pear and Apple (Fig. 196).

(e) *Aggregated Fruit*, a clustered and coherent mass of carpels, the product of a single flower, as in Raspberry (Fig. 197).

(f) *Multiple Fruit*, a clustered and coherent mass of carpels, each carpel being the product of a separate flower, as in Pine-apple. The cone of the Pine may be regarded as a *dry* multiple fruit (Fig. 198).

(g) *Accessory Fruit*, one in which the most conspicuous part is neither a part of the pistil nor combined with it. as in Strawberry, where the conspicuous part is only the enlarged and brightly coloured receptacle, the true fruit consisting of the achenes which dot its surface (Fig. 199), and in Sweet Brier, where the fleshy outer part is a calyx-tube lined with a hollow receptacle which bears the true fruit (achenes) on its inner surface (Fig. 200).

Fig. 196.

Fig. 197.

Fig. 198.

THE SEED.

Definition.

The seed is the mature ovule, and is specially characterized by the presence of the embryo or young plantlet.

Parts

INTEGUMENT: formed by the development of the coats of the ovule, and consisting of an outer and an inner layer.

(a) *Testa*, the outer layer (Fig. 206).

(b) *Tegmen*, the inner layer (Fig. 206).

In connection with the integument note

(1) *The Funiculus*, already defined when describing the ovule.

(2) *The Hilum*, or scar where the funiculus was attached.

(3) *The Micropyle*, a minute opening through the integument.

Also the following special appendages:

(1) *Aril*, an outgrowth of the funiculus or placenta, forming a more or less fleshy covering outside the true integument of certain seeds, as in the Climbing Bitter-Sweet and the White Water Lily (Fig. 201).

Fig. 199.

Fig. 200.

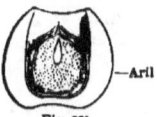
—Aril
Fig. 201.

(2) *Coma*, a tuft of hairs attached to the testa in some seeds, as in Willow-herb and Milk-weed (Fig. 202).

The coma must not be confounded with the *pappus* of composite flowers; the latter is attached to the *fruit*.

(3) *Wing*, a thin expansion of the testa (Fig. 203). But in the seeds of the Pine the wing splits off from the *scale* upon which the seed grows (Fig. 168).

NUCLEUS: the body of the seed within the integument, containing

(a) *Embryo*, the young plantlet as found in the seed. This is made up of

(1) *Radicle*, the rudimentary stem (Fig. 204).

(2) *Cotyledons*, or *Seed-leaves*, the first leaves, often thick and fleshy, as in the Bean (Fig. 205), but sometimes thin and leaf-like.

(3) *Plumule*, the bud at the top of the radicle (Fig. 204).

(b) *Albumen* or *Endosperm*, when present; nourishing matter stored up outside the embryo, as shown in the shaded portion of Fig. 206, the light part in the centre being the embryo.

Fig. 202.

Fig. 203.

cotyledon

plumule
radicle
Fig. 204.

cotyledon

Fig. 205.

Kind.

DICOTYLEDONOUS: having two cotyledons (Figs. 207, 208, 209).
MONOCOTYLEDONOUS: having only one cotyledon (Figs. 210, 211, 212).
POLYCOTYLEDONOUS: having several cotyledons (Fig. 206). This is rare.
ACOTYLEDONOUS: having no cotyledons (rare).
ALBUMINOUS: having albumen or endosperm in addition to the embryo (Figs. 206, 210).

The following terms apply to the folding of the parts of the embryo in dicotyledonous seeds:

(a) *Accumbent*, when the radicle is turned so as to touch the edges of the cotyledons (Fig. 213).

(b) *Incumbent*, when the radicle is turned so as to lie against the back of one cotyledon (Fig. 214).

(c) *Conduplicate*, the same as *incumbent* with the addition that the cotyledons are curved so as to partly infold the radicle (Fig. 215).

Nature and Use of the Parts of the Flower.

All the parts of the flower are leaf-forms (phyllomes), differing from ordinary foliage-leaves because their functions are different.

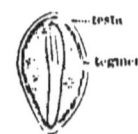
testa
tegmen
Fig. 206.

Figs. 207. 208. 209.

Fig. 210.

Fig. 211.

Fig. 212.

Fig. 213.

Fig. 214.

Fig. 215.

BOTANICAL TERMS.

The sepals differ less in appearance from ordinary leaves than any of the other parts. The petals resemble foliage-leaves in shape, but are mostly bright-coloured instead of green, and they are often sweet-scented. Sepals and petals together are *protective* organs, and they also serve to attract insects.

Stamens are leaf-forms in which the filament answers to the petiole, and the anther to the blade, as shown in Fig. 216.

Carpels are leaf-forms folded lengthwise more or less completely, as shown in Fig. 217.

Stamens and carpels are *essential* organs, and are directly concerned in the production of seed.

FLORAL DIAGRAMS.

By a floral diagram is meant the plan of a flower as exhibited in a cross-section. It should show the number and relative position of all the floral organs. The position of sepals, petals, and stamens is commonly easy to fix, but the true position of the carpels presents a little more difficulty. The ovary must be cut across with a sharp knife while some other organ (say the calyx) is still in position, and the relative situation of the carpels must then be carefully observed. The æstivation of calyx and corolla may also be shown to advantage in a floral diagram.

A number of examples of these diagrams are given in the margin, and the pupil should make the construction of such diagrams a regular part of his work.

Fig. 218 is a diagram of a Mint flower.
Fig. 219 " " " Leguminous flower.
Fig. 220 " " " Marsh Marigold.
Fig. 221 " " " Melon (staminate).
Fig. 222 " " " Melon (pistillate).
Fig. 223 " " " Composite flower.
Fig. 224 " " " Iris.
Fig. 225 " " " Grass flower.

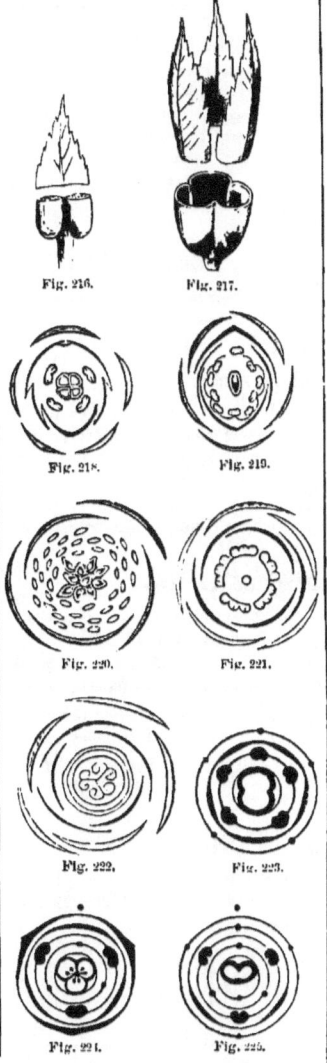

Fig. 216. Fig. 217.
Fig. 218. Fig. 219.
Fig. 220. Fig. 221.
Fig. 222. Fig. 223.
Fig. 224. Fig. 225.

LONGITUDINAL SECTIONS.

The pupil should make a constant practice of *splitting* flowers through the centre (best done from below upwards with a very sharp knife), and drawing the section thus presented. Such a drawing is exceedingly useful in connection with the floral diagram, as still further exhibiting the relation of the parts to each other.

Fig. 226 is a good example. Here the relations of the parts can be seen at a glance.

Fig. 226.

COMPOSITE FLOWERS.

A full description of a Composite flower involves some particulars of a special kind; for convenience, therefore, the various terms in use are collected together here.

Inflorescence.

Under this heading describe the arrangement of the heads, using the terms already explained—solitary, cymose, racemose, corymbose, spiked, etc.

Fig. 227.   Fig. 228.

Head.

The assemblage of florets (few or many) on a common receptacle.

Parts of the Head.

FLORETS: the small single flowers which in the aggregate make up the head. These are

(a) *Ligulate*, when the corolla is prolonged on one side into a flat strap-shaped piece (Fig. 227).

(b) *Tubular*, when the corolla is not thus prolonged, but is regularly developed all round (Fig. 228).

RECEPTACLE: the place upon which the florets stand.

INVOLUCRE: the circle or circles of bracts which surround the head.

Fig. 229.

Kinds of Head.

LIGULIFLORAL: when all the florets of the head are ligulate, as in Dandelion (Fig. 229).

TUBULIFLORAL: when all the florets are not ligulate; and such heads are

(a) *Discoid*, if all the florets are tubular as in Thistle.

(b) *Radiate*, if the florets round the margin of the head *ray-florets* are ligulate, while the central ones *disk-florets* are tubular, as in Sunflower (Fig. 230).

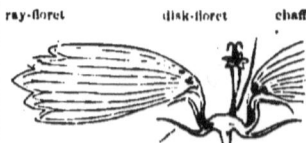
ray-floret   disk-floret   chaff
involucre   Fig. 230.

Ray-Florets (always without stamens).

NUMBER: 5, 10, 20. ∞, etc.

KIND:
    (a) *Pistillate*, if the pistil is present.
    (b) *Neutral*, if the pistil is absent.

SHAPE: linear, oblong, ovate, etc.

COLOUR: white, yellow, etc.

PAPPUS (if present):
    (a) *Simple*, if in a single row of similar pieces.
    (b) *Double*, if there is an outer row of shorter pieces.
    (c) *Capillary*, of fine hair-like pieces.
    (d) *Plumose*, of branching hairs or bristles, as in Thistle.
    (e) *Barbed*, if the hairs have teeth pointing backward, as in Dandelion.
    (f) *Chaffy*, of a few teeth or scales (Fig. 231).

Fig. 231.

ACHENE:
    (a) *Compressed*, when somewhat flattened.
    (b) *Terete*, cylindrical (the cross-section round).
    (c) *Angled*, as in Fig. 232.
    (d) *Striate*, marked with fine vertical lines.

Disk-Florets.

NUMBER: 5, 10, 20, ∞, etc.

KIND: perfect, staminate, etc.

COLOUR: yellow, brown, etc.

PAPPUS: as for the ray-florets.

ACHENE: as for the ray-florets.

Fig. 232.

Receptacle.

FORM: flat, concave, convex, conical, etc.

SURFACE:
    (a) *Chaffy*, if there are chaff-like scales or bristles growing on the receptacle among the florets (Fig. 230), as in Sunflower.
    (b) *Smooth*, or *naked*, if there are no such scales or bristles, as in Dandelion.

Involucre.

FORM:
    (a) *Ovoid*, egg-shaped, the broader part below, as in Thistle.
    (b) *Cylindrical*, nearly the same width all the way up (Fig. 233).
    (c) *Saucer-shaped*, very flat and shallow.
    (d) *Cup-shaped*, *Bell-shaped*, etc.

Fig. 233.

Bracts (or Scales) or Involucre.

NUMBER OF ROWS: state the exact number, unless very numerous.

ARRANGEMENT OF SCALES:

(a) *Imbricated*, in several rows and over-lapping (Fig. 233).

(b) *Reflexed*, turned backward, as in Dandelion (Fig. 229).

(c) *Appressed*, closely pressed together.

(d) *Squarrose*, with the points widely spreading (Fig. 233).

TEXTURE:

(a) *Herbaceous*, green and leaf-like.

(b) *Scarious*, thin and membranaceous.

SHAPE: use the ordinary leaf terms.

Fig. 234.

Fig. 235.

GRASSES.

These plants also require several special terms for their complete description. A few of the most necessary are given here.

Inflorescence.

In nearly all cases the inflorescence is a *panicle*, that is, an irregularly branched raceme, and the panicle is either loose and open, as in Meadow-grass (Fig. 238), or dense and closely packed as in Timothy and Foxtail.

SPIKELETS: the small separate clusters of flowers which together make up the panicle (Fig. 239). In some cases there is but *one* flower in the spikelet.

OUTER GLUMES: the pair of bracts at the base of the spikelet (Fig. 240). Note their shape and relative size.

INNER GLUMES OR PALEÆ: the pair of chaff-like bracts enclosing the particular flower (Fig. 242).

AWNS: bristle-like appendages sometimes found on the glumes or paleæ (Fig. 242).

LODICULES: small hypogynous scales next to the stamens, occasionally four (?) in grass-flowers.

Culm.

This is the name of the stem (Fig. 238). It is usually hollow except at the joints. The culms may be tufted or single, and their position and other characters can be described by terms already explained.

Fig. 237.   Fig. 236.

Leaf.

SHEATH: the lower portion of the leaf surrounding the stem, and split on the side away from the blade (Fig. 238).

LIGULE: a thin upward projection from the top of the sheath.

Fruit.

This always a *caryopsis* or *grain*.

TYPES OF GRASSES.

The following selection of Grasses will be found useful for examination, as illustrating most of the variations in the structure of these plants.

1. **Timothy.**
   Note the close inflorescence. Separate one of the component pieces which will probably resemble Fig. 234. If fully opened out it will resemble Fig. 235. Carefully dissect and describe, making a floral diagram. The spikelet here consists of a single flower.

2. **Red-Top.**
   Note the open panicle (Fig. 236). Detach and dissect a spikelet (Fig. 237), which in this plant also consists of a single flower. Observe the difference in the size of the inner bracts, and the three nerves on the larger one.

3. **Meadow-Grass.**
   The inflorescence is here an open greenish panicle, but each spikelet (Fig. 239) is compressed laterally and contains from three to five flowers. Fig. 240 shows a single flower. Note the delicate whitish margin of the lower palet, and the thin texture of the upper one; also the two teeth at the apex of the latter, and the five nerves on the former.

4. **Chess.**
   Here the spikelets (Fig. 241) are on long, slender, nodding pedicels, and each contains from eight to ten flowers. The glumes are different in size. Dissect out a single flower (Fig. 242) and note the awn on the lower palet. The upper palet at length grows fast to the groove of the oblong grain.

5. **Couch-Grass.**
   In this grass the spikelets are sessile on opposite sides of a zig-zag peduncle, so that the whole forms a sort of spike. Each spikelet is four to eight-flowered, and there is but one at each joint of the peduncle, the *side* of the spikelet being against the stalk. Note the running root-stocks, which cause the grass to be a nuisance difficult to get rid of.

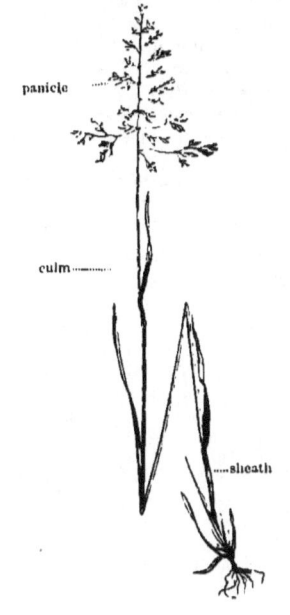

Fig. 238.

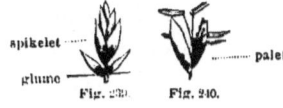

Fig. 239. Fig. 240.

Fig. 241. Fig. 242.

6. Old-Witch Grass.

This grass is to be found everywhere in sandy soil and in cultivated grounds. The leaves are very hairy, and the panicle very large, compound, and loose, the pedicels being extremely slender. Of the two glumes one is much larger than the other. Unless you are careful you will regard the spikelets as 1-flowered; observe, however, that in addition to the one manifestly perfect flower *there is an extra palet below*. This palet (which is very much like the larger glume) is a rudimentary or abortive second flower, and the spikelet may be described as 1½-flowered.

7. Barnyard Grass.

This is a stout, coarse plant, common in manured soil. It is from one to four feet in height, and branches from the base. The spikelets form dense spikes, and these are crowded in a dense panicle which is rough with stiff hairs. The structure of the spikelets is much the same as in Old-Witch Grass, but the palet of the neutral flower is pointed with a rough awn or bristle.

8. Foxtail.

In the common Foxtail the inflorescence is apparently a dense, bristly, cylindrical spike. In reality, however, it is a spiked panicle, the spikelets being much the same as in Barnyard Grass, but their *pedicels* are prolonged beyond them into awn-like bristles. In this plant the bristles are in clusters and are barbed upwards. *The spikes are twenty-y flow in colour.*

THE PLANT-BODY GENERALLY,
And the Functions of its Parts.

The higher plants, such as phanerogams, are found to be made up of four distinct kinds of members, as follows:

A.—ROOT, embracing the ordinary subterranean forms as previously described, and certain aerial forms, together with those of parasitic plants which feed upon other living organisms. The root differs from the stem in several important respects:

(a) It is tipped with a mass of hardened cells constituting the *root-cap* (Fig. 243, *a*). This protects the young root as it makes its way through the soil, and it is replaced from the inside as fast as it is worn away on the outside.

(b) The growth of the root in length is accomplished by additions to its extremity, immediately behind the root-cap.

Fig. 243.

(c) Roots originate *endogenously*, that is to say, they do not develope from the exterior or surface of the plant-body, but always begin in the deeper tissues, and eventually break their way through the overlying layers till they reach the surface.

(d) They do not, as a rule, produce leaves or buds.

(e) They tend, as a rule, to grow downwards into the soil, avoiding the light.

(f) The minute structure of the root is less perfect in its development than that of the stem.

The *functions* of the root are

(a) To fix the plant in its place.

(b) To act as an absorbent of the nutritious liquids contained in the soil.

(c) In special cases to serve as a storehouse of food for the plant.

B.—CAULOME: including the stem and all its equivalents, such as branches, runners, tendrils, thorns, etc., as already described.

In contrast to the root, the stem is always preceded by a *bud*.

A bud is an early stage of the development of a stem or branch, and is found on dissection to consist of many rudimentary leaves crowded on a short axis. This axis subsequently developes *throughout its length*, forming the internodes (Fig. 244), thus differing widely from the root, which grows by additions to its extremity.

Winter-buds are covered with scaly bracts called *bud-scales*, which separate and fall away soon after the development of the bud begins in the spring.

Buds are

(a) *Terminal*, when at the ends of stems and branches.

(b) *Axillary*, when produced in the axils (Fig. 244) of leaves.

(c) *Adventitious*, when produced in some irregular manner.

(d) *Accessory*, when produced as extra or additional buds beside the regular axillary bud, so that there are really several buds in the axil.

The *functions* of the caulome are

(a) To bear leaves and flowers.

(b) To serve as a medium for the conveyance of the nourishing liquids absorbed by the root.

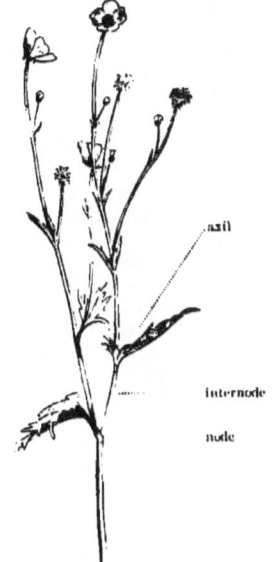

Fig. 244.

(c) In certain cases to serve as a storehouse for plant-food.

C. PHYLLOME: including the leaves and all their equivalents, such as bracts, cotyledons, bud-scales, sepals, petals, etc., as already described.

The phyllome is always developed laterally on a caulome.

Foliage-leaves (as contrasted with flower-leaves) are generally green, owing to the presence of a substance called *chlorophyll* (found also in all other green parts). A section through the body of a leaf is shown in Fig. 245, the shaded portions representing the cells which contain chlorophyll.

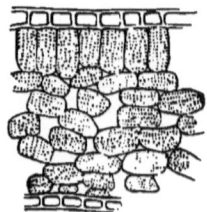

Fig. 245.

The chief *function* of foliage-leaves is to assimilate the food-materials derived from the soil and the air, thus converting them into forms (commonly starch) which can be used in advancing the plant's growth. Sunlight and chlorophyll are essential to the process of assimilation.

Transpiration. Water-vapour is given off through the leaves, by the agency of minute openings (chiefly on the under surface) known as *stomata* (singular *stoma*). One of these greatly magnified is shown in Fig. 246. These stomata communicate with air-spaces among the loosely-packed cells in the body of the leaf. It may often be observed in hot bright weather that the leaves of plants droop if exposed to the sun; this is because the loss of water through the leaves is greater than the supply through the roots. At night, however, the stomata close, and the balance being restored the plant recovers.

Fig. 246.

The functions of flower-leaves have already been referred to.

D. TRICHOME: including all the outgrowths from the surface or epidermis, whether of stem, leaf or root, such as hairs, bristles, root-hairs, prickles, etc.

Of all the trichome structures the root-hairs which occur abundantly on the young roots of most plants, are the most important. They consist of single long cells, and their function is to increase the absorbing surface of the root, for which service they are peculiarly fitted by the thin and delicate nature of their walls.

Fig. 247.

Hairs on parts above ground usually consist of a row of cells placed end to end (Figs. 217, 218). Often they are branched, as in the leaf-hairs of the Mullein.

Glandular hairs secrete a liquid in the cell which occupies the extremity of the hair. The sticky surfaces of certain plants are produced in this way.

Fig. 248.

Stinging hairs secrete a poisonous liquid. The point of a hair of this kind on piercing the skin breaks off, leaving the poison in the wound.

Prickles (Fig. 249) differ from *thorns* in being outgrowths of the bark; thorns arise from the wood.

GROWTH.

The growth of a plant consists in the multiplication of its cells, and the subsequent enlargement of the latter by the addition of new matter. The development of the cell frequently involves, also, a change of form.

Germination.

By this term is meant the commencement of the process of growth from the seed. Under suitable conditions of temperature and moisture the embryo, which is dormant in the dry seed, wakens into activity and begins to develope. The details of the process vary somewhat according to the structure of the seed. If the cotyledons are thin and leaf-like, as in Maple for example, the radicle generally grows throughout its length so as to raise them above the soil, where they at once expand and become the first green leaves of the new plant, a root being at the same time developed from the lower end of the radicle. But if the cotyledons are thick and fleshy, containing much nourishment, then usually a bud called the plumule, which contains the elements of additional bits of stem, will be a prominent feature in the embryo, and in this case the cotyledon or cotyledons not infrequently remain under ground, as in the pea and the acorn, and so do not perform the office of foliage-leaves, but merely supply the newly developing parts with nourishment. In albuminous seeds, the endosperm is the chief source from which the germinating embryo derives its support.

Vitality of Seeds

There is a considerable difference in regard to the length of time during which seeds retain their vitality. Some, such as those of Elm and Poplar, will germinate only if they have been kept fresh and not permitted to dry up, while others, such as those of Indian Corn and Wheat, and in general those containing a copious store of starch, may be kept for a very long time without losing their germinating power.

Food of Plants.

Growth implies assimilation of food. The elements of plant-food are ascertained by making a chemical analysis of the plant itself. Water forms a very considerable percentage of the whole weight, but is present to a greater extent in some portions of the plant body than in others. Fleshy roots, for example, may contain as much as 90 per cent., while dry seeds contain only about 12 per cent.

The water may be expelled by careful drying, and if what is then left is burnt, what is called the *organic* part of the plant disappears, and the *inorganic* part (the ash) remains behind. The organic part consists mainly of the elements carbon, hydrogen, oxygen, nitrogen, and sulphur; while the inorganic part contains very small quantities of phosphorus, iron, calcium, magnesium, and potassium. Of all these constituents of the *dry* plant carbon is the most abundant, amounting to about half the entire weight.

Sources of Plant-food.

All the materials just mentioned are obtained from the air, the water, and the soil. There is constantly present in the air carbonic acid gas—a compound of carbon and oxygen. This is absorbed by the leaves of land-plants, and (being soluble) from the water in which they live, by immersed plants. After absorption the gas is decomposed and the carbon appropriated. The oxygen required by the plant is derived chiefly from the carbonic acid gas and from water. Hydrogen is obtained chiefly by the decomposition of water, and nitrogen from the nitrates and ammonia salts in the soil. Sulphur, also, is obtained from salts occurring in the soil, and so too, of course, are all the inorganic elements

Respiration.

Plants, like animals, are continually inhaling oxygen; indeed, as with animals oxygen is essential to their existence. Germinating seeds and growing parts require large quantities of oxygen. The gas when inhaled is combined with carbon, giving rise to carbon dioxide. This process of oxidation is always accompanied by evolution of heat. This is well illustrated in the process of malting, where damp barley is heaped together. As soon as the grain begins to sprout oxygen is rapidly absorbed, and a very decided rise of temperature takes place.

Assimilation.

This is the process by which the carbon obtained from carbon dioxide is combined with the elements of water to form starch.

Metastasis.

This is the process by which the starch, resulting from assimilation, is converted into soluble forms and removed from the cells where it was produced to other portions of the plant where it is needed for purposes of growth, or, if there is an excess, to storehouses such as roots, bulbs, etc., for future use.

Circumstances Affecting Growth.

Temperature.—Growth may be stopped altogether by either too low or too high a temperature, and between the limits within which any given plant is found to be capable of growth there will be found a particular degree of temperature more favourable to growth than any other, either above it or below it. This may be called the *optimum*. The effect of temperature differs considerably according to the amount of water present in the part affected, dry seeds, for instance, resisting a temperature, either high or low, to which soaked seeds would at once succumb.

Light.—Light is essential to assimilation, but seeds and tubers, as well as many of the lower plants which are without chlorophyll, such as Mushrooms, will grow in the absence of light as long as the stock of assimilated material upon which they draw is not exhausted. The growth which takes place in the cambium-layer of dicotyledons and in roots is another example of increase in size in the absence of light. The assimilated material in all these cases, however, has been previously elaborated elsewhere.

Light is found to exercise a retarding influence upon growth. A plant, for instance, in a window will bend towards the light, because the cells on the side nearest the window grow more slowly than those which are shaded, thus causing curvature of the stem and petioles.

Gravitation.—Gravitation also affects growth, as we know that the stem and root, or *axis* of the plant, are usually in the line of the radius of the earth at the place of growth. If a seedling plantlet be laid with the stem and root horizontal, the stem will curve upward and the root downward in the endeavour to restore the vertical direction.

THE HERBARIUM.

Those who are anxious to make the most of their botanical studies will find it of great advantage to gather and preserve specimens for reference. A few hints, therefore, on this subject will not be out of place. It will, of course, be an object to collectors to have their specimens exhibit as many of their natural characters as possible, so that, although dried and pressed, there will be no difficulty in recognizing them; and to this end neatness and care are the first requisites.

Collecting.

Specimens should be collected when the plants are in flower, and, if possible, on a dry day, as the flowers are then in better condition than if wet. If the plant is small, the whole of it, root and all, should be taken up; if too large to be treated in this way, a flower and one or two of the leaves (radical as well as cauline, if these be different) may be gathered.

Drying.

As many of your specimens will be collected at a distance from home, a close tin box, which may be slung over the shoulder by a strap, should be provided, in which the plants may be kept fresh, particularly if a few drops of water be sprinkled upon them. Perhaps a better way, however, is to carry a portfolio of convenient size—say 15 inches by 10 inches—made of two pieces of stout pasteboard or thin deal, and having a couple of straps with buckles for fastening it together. Between the covers should be placed sheets of blotting-paper or coarse wrapping-paper, as many as will allow the specimens to be separated by at least five or six sheets. The advantage of the portfolio is, that the plants may be placed between the sheets of blotting-paper, and subjected to pressure by means of the straps as soon as they are gathered. If carried in a box, they should be transferred to paper as soon as possible. The specimens should be spread out with great care, and the crumpling and doubling of leaves guarded against. The only way to prevent moulding is to place plenty of paper between the plants, and *change the paper frequently*; the frequency depending on the amount of moisture contained in the specimens. From ten days to a fortnight will be found sufficient for the thorough drying of almost any plant you are likely to meet with. Having made a pile of specimens with paper between them, as directed, they should be placed on a table or floor, covered by a flat board, and subjected to pressure by placing weights on the top; twenty bricks or so will answer very well.

It is of great importance that *the sheet of paper within which the plant is first placed* should not be interfered with during the drying process. The directions as to frequent changes refer only to the sheets not immediately in contact with the plant. These, to ensure the best results, should be changed once a day for the first few days; less frequently thereafter. Gray recommends ironing with hot irons in order to remove more rapidly the moisture from fleshy leaves, and in any case to warm the driers in the sun before putting them between the plants.

Mounting.

When the specimens are thoroughly dry, the next thing is to mount them, and for this purpose you will require sheets of strong white paper; a good quality of unruled foolscap or cheap drawing paper will be suitable. The most convenient way of attaching the specimen to the paper is to take a sheet of the same size as your paper, lay the specimen carefully in the centre, wrong side up, and gum it thoroughly with a very soft brush. Then take the paper to which the plant is to be attached, and lay it carefully on the specimen. You can then lift paper and specimen together, and, by pressing lightly with a soft cloth, ensure complete adhesion. To render plants with stout stems additionally secure, make a slit with a penknife through the paper immediately underneath the stem, then pass a narrow band of paper round the stem, and thrust both ends of the band through the slit. The ends may then be gummed to the back of the sheet.

Sorting and Ticketing.

The specimen having been duly mounted, its botanical name should be written neatly in the lower right-hand corner, together with the date of its collection and the locality where found. Of course only one Species should be mounted on each sheet; and when a sufficient number have been prepared, the Species of the same Genus should be placed in a sheet of larger and coarser paper than that on which the specimens are mounted, and the name of the Genus should be written outside on the lower corner. Then the Genera of the same Order should be collected in the same manner, and the name of the Order written outside as before. The Orders may then be arranged in accordance with the classification you may be using, and carefully laid away in a dry place. If a cabinet, with shelves or drawers, can be specially devoted to storing the plants, so much the better.

KEY TO THE FAMILIES OR ORDERS.

SERIES I. PHANEROGAMS.
Plants producing true flowers and seeds.

CLASS I. DICOTYLEDONS.
Distinguished ordinarily by having net-veined leaves, and the parts of the flowers in fours or fives, very rarely in sixes. Wood growing in rings, and surrounded by a true bark. Cotyledons of the embryo mostly two.

SUB-CLASS I. ANGIOSPERMS.
Seeds enclosed in an ovary.

1. POLYPETALOUS DIVISION.
Two distinct sets of Floral Envelopes. Parts of the corolla separate from each other.

A. Stamens more than twice as many as the petals.

* *Stamens hypogynous (inserted on the receptacle).*

+ *Pistil apocarpous (carpels separate from each other).*

RANUNCULACEÆ.—Herbs. Leaves generally decompound or much dissected............ 2
ANONACEÆ. Small trees. Leaves entire. Petals 6, in 2 sets 7
MAGNOLIACEÆ. Trees. Leaves truncate. Fruit resembling a cone.................. 6
MENISPERMACEÆ.- Woody twiners. Flowers dioecious. Leaves peltate near the edge........ 7

Brasenia, in
NYMPHÆACEÆ.- Aquatic. Leaves oval, peltate; the petiole attached to the centre........... 9
MALVACEÆ. Stamens monadelphous. Calyx persistent. Ovaries in a ring 24

Podophyllum, in
BERBERIDACEÆ.—Calyx fugacious. Leaves large, peltate, deeply lobed. Fruit a large fleshy berry, 1 celled............................ 8

+ + *Pistil apocarpous. (Stigmas, styles, placentæ, or cells, more than one.)*

Actæa, in
RANUNCULACEÆ, might be looked for here. Fruit a many-seeded berry. Leaves compound 2
NYMPHÆACEÆ. Aquatics. Leaves floating, large, deeply cordate.................. 9
SARRACENIACEÆ.—Bog-plants. Leaves pitcher-shaped................ 10
PAPAVERACEÆ.—Juice red or yellow. Sepals 2, caducous 10

CAPPARIDACEÆ.—Corolla cruciform, but pod 1-celled. Leaves of 3 leaflets 16
HYPERICACEÆ.—Leaves transparent-dotted. Stamens usually in 3, but sometimes 5, clusters.. 19
CISTACEÆ.—Sepals 5, very unequal, or only 3. Ovary 1 celled, with 3 parietal placentæ . . 18
MALVACEÆ.—Stamens monadelphous, connected with the bottom of the petals. Calyx persistent. Ovaries in a ring................. 24
TILIACEÆ.—Trees. Flowers yellowish, in small hanging cymes, the peduncle with a leaf-like bract attached............................. 25

* * *Stamens perigynous (inserted on the calyx).*

Portulaca, in
PORTULACACEÆ.—Low herbs, with fleshy leaves. Sepals 2, adhering to the ovary beneath. Pod opening by a lid...... 23
ROSACEÆ.—Leaves alternate, with stipules. Fruit apocarpous, or a drupe, or a pome.......... 38

* * * *Stamens epigynous (attached to the ovary).*

Nymphæa, in
NYMPHÆACEÆ. — Aquatic. Leaves floating. Flowers white, large, with numerous petals gradually passing into stamens 9

B. Stamens not more than twice as many as the petals.

* *Stamens just as many as the petals, and one stamen in front of each petal.*

BERBERIDACEÆ.—Herbs (with us). Anthers opening by uplifting valves 8
PORTULACACEÆ.—Sepals 2. Styles 3-cleft. Leaves 2, fleshy 23
VITACEÆ.—Shrubs, climbing by tendrils. Calyx minute. 29
RHAMNACEÆ.—Shrubs, not climbing 29

Lysimachia, in
PRIMULACEÆ, is occasionally polypetalous. Flowers yellow, in axillary spikes; the petals sprinkled with purplish dots 91

* * *Stamens either just as many as the petals and alternate with them, or not exactly the same number.*

+ *Corolla irregular.*

FUMARIACEÆ.—Corolla flattened and closed. Stamens 6....................... 11

KEY TO THE FAMILIES OR ORDERS.

VIOLACEÆ.—Corolla 1-spurred. Stamens 5. Pod with 3 rows of seeds on the walls........... 17

BALSAMINACEÆ.—Corolla 1-spurred, the spur with a tail. Stamens 5. Pod bursting elastically. 27

POLYGALACEÆ.—Lower petal keel-shaped, usually fringed at the top. Anthers 6 or 8, 1-celled, opening at the top. Pod 2-celled........... 32

LEGUMINOSÆ.—Corolla mostly papilionaceous. Filaments often united. Ovary simple, with one parietal placenta. Leaves compound.... 33

++ *Corolla regular, or nearly so.*

1. **Calyx superior** (*i.e.,* adherent to the ovary, wholly or partially).

   (a) Stamens perigynous (inserted on the calyx).

Cratægus, in

ROSACEÆ.—Shrubs. Stamens occasionally from 5 to 10 only. Leaves alternate, with stipules. Fruit drupe-like, containing 1-5 bony nutlets. 38

SAXIFRAGACEÆ.—Leaves opposite or alternate, without stipules. Styles or stigmas 2; in one instance 4. Ovary 1-celled, with 2 or 3 parietal placentæ...................... 46

HAMAMELACEÆ.—Shrubs. Stamens 8; styles 2. Flowers yellow, in autumn................ 48

HALORAGEÆ.—Aquatics. Stamens 4 or 8. Styles or sessile stigmas 4................... 49

ONAGRACEÆ.—Flowers symmetrical. Stamens 2, 4, or 8. Stigmas 2 or 4, or capitate........ 49

MELASTOMACEÆ.— 1-celled, opening by a pore at the stamens 8. Style and stigma 1. rple................. 51

LYTHRACEÆ.— rently adherent to, but really fr ovary. Stamens 10, in 2 sets. I whorled............. 51

CUCURBITA' ll-bearing herbs. Flowers monœc 52

(b) Stame (*on the ovary, or on a disk which ers the ovary*).

Euonymu,

CELASTRACEÆ —shrub, with 4-sided branchlets, not climbing. Leaves simple. Pods crimson when ripe. Calyx not minute............. 30

UMBELLIFERÆ.—Flowers chiefly in compound umbels. Calyx very minute. Stamens 5. Styles 2. Fruit dry, 2-seeded.............. 53

ARALIACEÆ.—Umbels not compound, but sometimes panicled. Stamens 5. Styles usually more than 2. Fruit berry-like............. 56

CORNACEÆ.—Flowers in cymes or heads. Stamens 4. Style 1..................... 57

2. **Calyx inferior** (*i.e.,* free from the ovary).

   (a) Stamens hypogynous (on the receptacle).

CRUCIFERÆ.—Petals 4. Stamens 6, tetradynamous. Pod 2-celled........................ 12

CISTACEÆ.—Petals 3. Sepals 5, very unequal; or only 3. Pod partly 3-celled.............. 18

DROSERACEÆ.—Leaves radical, beset with reddish glandular hairs. Flowers in a 1-sided raceme 19

Elodes, in

HYPERICACEÆ.—Leaves with transparent dots. Stamens 9, in 3 clusters................ 19

CARYOPHYLLACEÆ.—Styles 2-5. Ovules in the centre or bottom of the cell. Stem usually swollen at the joints. Leaves opposite...... 21

LINACEÆ.—Stamens 5, united below. Pod 10-celled, 10-seeded..................... 25

GERANIACEÆ.—Stamens 5. Carpels 5,—they and the lower parts of the 5 styles attached to a long beak, and curling upwards in fruit..... 26

OXALIDACEÆ.—Stamens 10. Pod 5-celled. Styles 5, distinct. Leaflets 3, obcordate, drooping at night-fall......................... 27

ERICACEÆ.—Anthers opening by pores at the top, or across the top. Leaves mostly evergreen, sometimes brown beneath; but in some instances the plant is white or tawny......... 85

(b) Stamens perigynous (plainly attached to the calyx).

SAXIFRAGACEÆ.—Leaves opposite or alternate, without stipules. Styles or stigmas 2; in one instance 4. Carpels fewer than the petals... 46

CRASSULACEÆ.—Flowers *symmetrical.* Stamens 10 or 8. Leaves sometimes fleshy............ 48

LYTHRACEÆ.—Stamens 10, in two sets. Calyx enclosing, but really free from, the ovary. Leaves mostly whorled................. 51

(c) Stamens attached to a fleshy disk in the bottom of the calyx-tube.

ANACARDIACEÆ.—Trees, or shrubs, not prickly. Leaves compound. Stigmas 3. Fruit a 1-seeded drupelet..................... 28

CELASTRACEÆ.—Twining shrub. Leaves simple. Pods orange when ripe.................. 30

SAPINDACEÆ.—Shrubs, or trees. Fruit 2-winged, and leaves palmately-veined. Or, Fruit an inflated 3-celled pod, and leaves of 3 leaflets. Styles 2 or 3......................... 31

(d) Stamens attached to the petals at their very base.

Claytonia, in

PORTULACACEÆ.—Sepals 2. Leaves fleshy. Style 3-cleft............................ 23

AQUIFOLIACEÆ.—Shrubs, with small axillary flowers, having the parts in fours or sixes. Fruit a red berry-like drupe. Stigma sessile. Calyx minute........................ 90

II. GAMOPETALOUS DIVISION.

Corolla with the petals united together, in however slight a degree.

KEY TO THE FAMILIES OR ORDERS.

A. Calyx superior (adherent to the ovary).

* *Stamens united by their anthers.*

CUCURBITACEÆ.—Tendril-bearing herbs 52
COMPOSITÆ.—Flowers in heads, surrounded by an involucre 64
LOBELIACEÆ.—Flowers not in heads. Corolla split down one side 83

* * *Stamens not united together in any way.*

+ *Stamens inserted on the corolla.*

DIPSACEÆ. Flowers in heads, surrounded by an involucre. Plant prickly 63
VALERIANACEÆ.—Flowers white, in clustered cymes. Stamens fewer than the lobes of the corolla 63
RUBIACEÆ.—Leaves, when opposite, with stipules; when whorled, without stipules. Flowers, if in heads, without an involucre 61
CAPRIFOLIACEÆ. Leaves opposite, without stipules; but, in one genus, with appendages resembling stipules 58

+ + *Stamens not inserted on the corolla.*

CAMPANULACEÆ. Herbs with milky juice. Stamens as many as the lobes of the corolla .. 63
ERICACEÆ.—Chiefly shrubby plants or parasites. Stamens twice as many as the lobes of the corolla 85

B. Calyx inferior (free from the ovary).

* *Stamens more than the lobes of the corolla.*

LEGUMINOSÆ.—Ovary 1-celled, with 1 parietal placenta. Stamens mostly diadelphous 33

Adlumia, in

FUMARIACEÆ.—Plant climbing. Corolla 2-spurred. 11
MALVACEÆ. Filaments monadelphous. Carpels in a ring 24
ERICACEÆ. Chiefly shrubby plants, with simple entire leaves. Stamens twice as many as the lobes of the corolla 85
POLYGALACEÆ. Anthers 6 or 8, 1-celled, opening at the top. Pod 2-celled. Flowers irregular; lower petal keel-shaped, and usually fringed at the top 32
OXALIDACEÆ. Stamens 10, 5 of them longer. Styles 5, distinct. Leaflets 3, obcordate, drooping at night-fall. 27

* * *Stamens just as many as the lobes of the corolla, one in front of each lobe.*

PRIMULACEÆ. Stamens on the corolla. Ovary 1-celled, with a free central placenta rising from the base 91

* * * *Stamens just as many as the lobes of the corolla, inserted on its tube alternately with its lobes.*

+ *Ovaries 2, separate.*

APOCYNACEÆ. Plants with milky juice. Anthers converging round the stigmas, but not adherent to them. Filaments distinct 114
ASCLEPIADACEÆ.—Plants with milky juice. Anthers adhering to the stigmas. Filaments monadelphous. Flowers in umbels 114

+ + *Ovary 4-lobed around the base of the style.*

Mentha, in

LABIATÆ. Stamens 4. Leaves opposite, aromatic 100
BORRAGINACEÆ.—Stamens 5. Leaves alternate... 105

+ + + *Ovary 1-celled ; the seeds on the walls.*

HYDROPHYLLACEÆ.—Stamens 5, usually exserted. Style 2-cleft. Leaves lobed and sometimes cut-toothed 108
GENTIANACEÆ.—Leaves entire and opposite ; or (in Menyanthes) of 3 leaflets 112

+ + + + *Ovary with 2 or more cells.*

AQUIFOLIACEÆ. Shrubs. Corolla almost polypetalous. Calyx minute. Fruit a red berry-like drupe. Parts of the flower chiefly in fours or sixes 90
PLANTAGINACEÆ.—Stamens 4. Pod 2-celled. Flowers in a close spike 91

Verbascum, in

SCROPHULARIACEÆ.—Corolla nearly regular. Flowers in a long terminal spike. Stamens 5; the filaments, or some of them, woolly 94
POLEMONIACEÆ.—Style 3-cleft. Corolla salver-shaped, with a long tube. Pod 3-celled, few-seeded ; seeds small 109
CONVOLVULACEÆ.—Style 2-cleft. Pod 2-celled, generally 4-seeded ; seeds large. Chiefly twining or trailing plants 109
SOLANACEÆ.—Style single. Pod or berry 2-celled, many-seeded 110

* * * * *Stamens fewer than the lobes of the corolla ; the corolla mostly irregular or 2-lipped.*

LABIATÆ.—Ovary 4-lobed around the base of the style. Stamens 4 and didynamous, or occasionally only 2 with anthers. Stem square .. 100
VERBENACEÆ.—Ovary 4-celled, but not lobed ; the style rising from the apex. Or, ovary 1-celled and 1-seeded. Stamens didynamous 99
LENTIBULACEÆ.—Aquatic. Stamens 2. Ovary 1-celled, with a free central placenta 93
OROBANCHACEÆ. Parasitic herbs, without green foliage. Ovary 1-celled, with many seeds on the walls. Stamens didynamous 94
SCROPHULARIACEÆ. Ovary 2-celled, with many seeds. Stamens didynamous, or only 2..... 94

KEY TO THE FAMILIES OR ORDERS.

III. APETALOUS DIVISION.

Corolla (and sometimes calyx also) wanting.

A. Flowers not in catkins.

* *Calyx and corolla both wanting.*

SAURURACEÆ.—Flowers white, in a dense terminal spike, nodding at the end. Carpels 6 or 4, nearly separate........................ 124

CERATOPHYLLACEÆ. — Immersed aquatics, with whorled finely dissected leaves. Flowers monœcious 124

* * *Calyx superior (i.e., adherent to the ovary).*

SAXIFRAGACEÆ.—Small, smooth herbs, with inconspicuous greenish-yellow flowers. Stamens twice as many as the calyx-lobes, on a conspicuous disk...... 46

HALORAGEÆ.—Aquatics. Leaves finely dissected or linear. Stamens 1-8. Ovary 4-lobed or (Hippuris) 1-celled........................ 49

ONAGRACEÆ.—Herbs, in ditches. Stamens 4. Ovary 4-celled, 4-sided.................. 49

ARISTOLOCHIACEÆ.—Calyx 3-lobed, dull purple inside. Ovary 6-celled................ 116

SANTALACEÆ.—Low plants with greenish-white flowers in terminal clusters. Calyx-tube prolonged, and forming a neck to the 1-celled nut-like fruit.......................... 124

ELÆAGNACEÆ. — Shrubs with scurfy leaves. Flowers diœcious. Calyx 4-parted, in the fertile flowers apparently adherent to the ovary, and becoming fleshy in fruit......... 123

* * * *Calyx inferior (plainly free from the ovary).*

+ *Ovaries more than one and separate from each other.*

RANUNCULACEÆ. — Calyx present, colored and petal-like. Achenes containing several seeds, or only one 2

RUTACEÆ.—Prickly shrubs, with compound transparent-dotted leaves, and diœcious flowers... 27

+ + *Ovary only one, but with more than one cell.*

CRASSULACEÆ.—Herbs, in wet places. Pod 5-celled and 5-horned................. 48

PHYTOLACCACEÆ.—Herbs. Ovary 10-celled and 10-seeded........................... 116

EUPHORBIACEÆ.—Herbs. Ovary 3-celled, 3-lobed, protruded on a long pedicel. Juice milky .. 125

SAPINDACEÆ.—Trees. Ovary 2-celled and 2-lobed. Fruit two 1-seeded samaras joined together. Flowers polygamous..................... 31

RHAMNACEÆ.—Shrubs. Ovary 3-celled and 3-seeded ; forming a berry 29

FICOIDEÆ.—Prostrate herbs with whorled leaves. Ovary 3-celled, many-seeded............... 52

URTICACEÆ.—Trees. Leaves simple. Ovary 2-celled, but fruit a 1-seeded samara winged all round. Stigmas 2................... .. 127

+ + + *Ovary only one, 1-celled and 1-seeded.*

POLYGONACEÆ.—Herbs. Stipules sheathing the stem at the nodes..................... 119

URTICACEÆ.—Herbs. Stigma 1. Flowers monœcious or diœcious, in spikes or racemes. No chaff-like bracts among the flowers. Or, Stigmas 2 ; leaves palmately-compound...... 127

AMARANTACEÆ. — Herbs. Flowers greenish or reddish, in spikes, *with chaff-like bracts interspersed.* Stigmas 2............ 118

CHENOPODIACEÆ.—Herbs. Flowers greenish, in spikes. *No chaff-like bracts.* Stigmas 2..... 116

OLEACEÆ.—Trees. *Leaves pinnately-compound.* Fruit a 1-seeded samara..................... 115

URTICACEÆ.—Trees. *Leaves simple.* Fruit a 1-seeded samara winged all round, or a drupe. 127

LAURACEÆ.—Trees or shrubs. Flowers diœcious. Sepals 6, petal-like. Stamens 9, opening by uplifting valves 122

THYMELEACEÆ.—Shrubs with leather-like bark, and jointed branchlets. Flowers perfect, preceding the leaves. Style thread-like 123

B. Flowers in catkins.

* *Sterile or staminate flowers only in catkins.*

JUGLANDACEÆ.—Trees with pinnate leaves. Fruit a nut with a husk........................ 130

CUPULIFERÆ.—Trees with simple leaves. Fruit one or more nuts surrounded by an involucre which forms a scaly cup or bur.............. 131

* * *Both sterile and fertile flowers in catkins, or catkin-like heads.*

SALICACEÆ.—Shrubs or low trees. Ovary 1-celled, many-seeded ; seeds tufted with down at one end 136

PLATANACEÆ.—Large trees. *Stipules sheathing the branchlets.* The flowers in heads 130

MYRICACEÆ.—Shrubs with resinous-dotted, usually fragrant, leaves. Fertile flowers one under each scale. Nutlets usually coated with waxy grains 134

BETULACEÆ.—Trees or shrubs. Fertile flowers 2 or 3 under each scale of the catkin. Stigmas 2, long and slender 135

SUB-CLASS II. GYMNOSPERMS.

Ovules and seeds naked, on the inner face of an open scale ; or, in Taxus, without any scale, but surrounded by a ring-like disk which becomes red and berry-like in fruit.

CONIFERÆ.—Trees or shrubs, with resinous juice, and mostly awl-shaped or needle-shaped leaves. Fruit a cone, or occasionally berry-like 139

KEY TO THE FAMILIES OR ORDERS.

CLASS II. MONOCOTYLEDONS.

Distinguished ordinarily by having straight-veined leaves (though occasionally net-veined ones), and the parts of the flowers in threes, never in fives. Wood never forming rings, but interspersed in separate bundles throughout the stem. Cotyledon only 1.

I. SPADICEOUS DIVISION.

Flowers collected on a spadix, with or without a spathe or sheathing bract. Leaves sometimes net-veined.

ARACEÆ.—Herbs (either flag-like marsh-plants, or terrestrial,) with pungent juice, and simple or compound leaves, these sometimes net-veined. Spadix usually (but not always) accompanied by a spathe. Flowers either without a perianth of any kind, or with 4-6 sepals 143

TYPHACEÆ.—Aquatic or marsh plants, with linear straight-veined leaves erect or floating, and monœcious flowers. Heads of flowers cylindrical or globular, no spathe, and no floral envelopes... 144

LEMNACEÆ.—Small aquatics, freely floating about 144

NAIADACEÆ.—Immersed aquatics. Stems branching and leafy. Flowers perfect, in spikes, generally on the surface... 145

II. PETALOIDEOUS DIVISION.

Flowers not collected on a spadix, furnished with a corolla-like, or occasionally herbaceous, perianth.

A. Perianth superior (adherent to the ovary).

* Flowers diœcious or polygamous, regular.

HYDROCHARIDACEÆ.—Aquatics. Pistillate flowers only above water; perianth of 6 pieces... 148

DIOSCOREACEÆ.—Twiners, from knotted root-stocks. Leaves heart-shaped, net-veined. Pod with 3 large wings... 157

** Flowers perfect.

ORCHIDACEÆ.—Stamens 1 or 2, gynandrous, Flowers irregular... 149

IRIDACEÆ. Stamens 3 155

AMARYLLIDACEÆ. Stamens 6. Flowers on a scape from a bulb... 156

B. Perianth inferior (free from the ovary).

ALISMACEÆ.—Pistil apocarpous; carpels in a ring or head, leaves with distinct petiole and blade 147

SMILACEÆ.—Climbing plants, with alternate ribbed and net-veined petioled leaves. Flowers diœcious ... 157

Triglochin, in

ALISMACEÆ.—Rush-like marsh herbs. Flowers in a spike or raceme. Carpels when ripe splitting away from a persistent axis ... 147

LILIACEÆ.—Perianth of similar divisions or lobes, mostly 6, but in one case 4. One stamen in front of each division, the stamens similar ... 158

Trillium, in

LILIACEÆ.—Perianth of 3 green sepals and three colored petals ... 158

PONTEDERIACEÆ.—Stamens 6, 3 long and 3 short. Perianth (blue or yellow) tubular, of 6 lobes. Aquatics ... 164

JUNCACEÆ.—Perianth glumaceous, of similar pieces 162

ERIOCAULONACEÆ.—In shallow water. Flowers in a small woolly head, at the summit of a 7-angled scape. Leaves in a tuft at the base... 165

III. GLUMACEOUS DIVISION.

Flowers without a true perianth, but subtended by thin scales called glumes.

CYPERACEÆ.—Sheaths of the leaves not split 165

GRAMINEÆ.—Sheaths of the leaves split on the side away from the blade... 168

SERIES II. CRYPTOGAMS.

Plants without stamens and pistils, reproducing themselves by spores instead of seeds.

CLASS III. PTERIDOPHYTES.

Stems containing vascular as well as cellular tissue.

FILICES.—Spores produced on the fronds ... 174

EQUISETACEÆ.—Spores produced on the under side of the shield-shaped scales of a terminal spike or cone... 181

LYCOPODIACEÆ.—Spore-cases produced in the axils of the simple leaves or bracts 182

ILLUSTRATIVE EXAMPLES

OF

PLANT DESCRIPTION.

A few examples of the method of filling plant schedules are given in the pages which immediately follow. They are intended to be suggestive rather than to be implicitly followed. Teachers will use their own judgment as to the degree of elaboration which will be aimed at in any particular case, as a good deal must depend upon the stage of the pupils' knowledge.

PLANT SCHEDULE.   NO. ...

ROOT.	Origin........	*Primary.*	**LEAF.** Division........*Simple.*
	Form........	*Tap, long and stout.*	Position........*Cauline.*
	Colour........	*White or whitish.*	Arrangement....*Alternate.*
	Duration....	*Biennial or perennial.*	Stipulation.....*Stipulate.*
	Position.....	*Subterranean.*	Insertion........*Petiolate; petioles very long.*
STEM.	Class........	*Dicotyledonous.*	Outline........*Round-kidney-shaped.*
	Attitude....	*Ascending or procumbent.*	No. of leaflets, if any...*None.*
	Texture.....	*Herbaceous.*	*Texture........*Thickish.*
	Position.....	*Aerial.*	*Colour........*Green both sides.*
	Shape.......	*Cylindrical.*	*Size..........*One to three inches across.*
	Juice........	*Mucilaginous.*	*Venation.....*Palmately net-veined*
	Branching...	*Stems simple, often tufted.*	*Margin.......*Slightly lobed and crenate.*
	Height......	*One to two feet.*	*Apex.........*Obtuse.*
	Duration...	*Dying to the ground annually.*	*Base.........*Deeply cordate.*
	Surface.....	*Pubescent.*	*Surface......*Pubescent.*

* Applicable to leaflets if leaf is compound.

INFLORESCENCE. Mode........*Racemose*   Variety........*One or two flowers in each axil.*

THE FLOWER.

ORGAN.	No.	COHESION.	ADHESION.	NOTES ON FORM, ÆSTIVATION, COLOUR, ETC.
		NOTE. This space need not be used except for Monocotyledons.		
Perianth. *Leaves.*				
Calyx. *Sepals.*	5	*Gamosepalous.*	*Inferior.*	*An epicalyx of three bracts. Calyx valvate, persistent.*
Corolla. *Petals.*		*Polypetalous.*	*Hypogynous.*	*Petals white or pinkish, obcordate, ½ inch long, convolute in the bud.*
Stamens. *Filaments. Anthers.*	∞	*Monadelphous.*	*Hypogynous.*	*Tube of stamens united with the base of the corolla. Anthers 1-celled.*
Pistil. *Stigma. Styles. Carpels. Ovary of.*		*Syncarpous.*	*Superior.*	*Carpels in a ring, as many as the styles, 1-seeded.*
FRUIT. Kind		*Dry, indehiscent.*		
Variety		*Schizocarp, breaking up into 1-seeded closed carpels.*		
Dehiscence		*Indehiscent.*		
No. of Seeds		*As many as the carpels.*		
Description of Seed		*Kidney shaped, cotyledons crumpled, little albumen.*		

FLORAL DIAGRAM.

CLASSIFICATION, &c.

SERIES	Phanerogams.	Botanical Name	Malva rotundifolia
CLASS	Angiosperms.	Popular Name	Round-leaved Mallow.
SUB-CLASS	Dicotyledons.	Habitat	Roadsides and cultivated soil.
DIVISION	Polypetalous.	Where found	Roadside, North Toronto.
Order	Malvaceæ.	Date of collection	September 20th, 1894.
Genus	Malva.		
Species	Rotundifolia.		

DRAWINGS, &C.

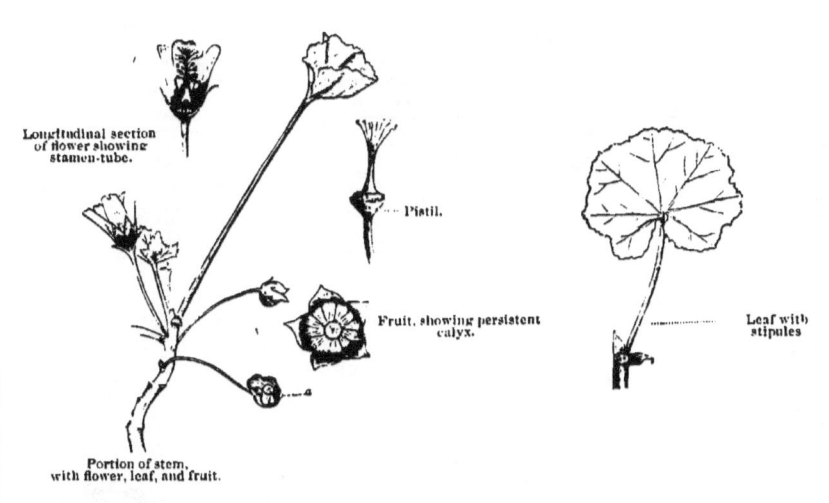

Longitudinal section of flower showing stamen-tube.

Pistil.

Fruit, showing persistent calyx.

Leaf with stipules.

Portion of stem, with flower, leaf, and fruit.

PLANT SCHEDULE.   NO. ....

ROOT.	Origin......*Secondary.*		**LEAF.**	Division......*Simple.*
	Form......*Fibrous.*			Position......*Radical.*
	Colour......*Whitish.*			Arrangement...*Alternate.*
	Duration....*Perennial.*			Stipulation.....*Exstipulate.*
	Position......*Subterranean.*			Insertion.......*Petioles sheathing the scape.*
STEM.	Class........*Monocotyledonous; a bulb.*			Outline........*Oblong-lanceolate.*
	Attitude....			No. of leaflets, if any..*None.*
	Texture.....*Herbaceous.*			*Texture........*Thickish and soft.*
	Position.....*Deep in the ground.*			*Colour.....*Green, mottled with purple above.*
	Shape........*Mostly oblong; small.*			*Size..........*Three to five inches long.*
	Juice........*Colourless.*			*Venation.....*Straight-veined.*
	Branching...*None.*			*Margin........*Entire.*
	Height......			*Apex..........*Acute.*
	Duration....*Perennial.*			*Base..........*Tapering.*
	Surface.....			*Surface.......*Smooth and shining.*
				* Applicable to leaflets if leaf is compound.
INFLORESCENCE.	Mode.........*Terminal.*			Variety........*Solitary.*

THE FLOWER.

ORGAN.	No.	COHESION.	ADHESION.	NOTES ON FORM, ÆSTIVATION, COLOUR, ETC.
Perianth. Leaves.	6	*Polyphyllous.*	*Inferior.*	Divisions spreading, lanceolate, yellow, purple-spotted, an inch long.
Calyx. Sepals.				
Corolla. Petals.				
Stamens. Filament. Anthers.	6 6 6	*Hexandrous.*	*Hypogynous.*	Stamens opposite the divisions of the perianth.
Pistil. Stigmas. Styles. Carpels. Ovary-cells.	1 1 3 3	*Syncarpous.*	*Superior.*	Ovary narrowed at the base. Style club-shaped. Stigma 3-lobed.
FRUIT.	Kind........*Dry; dehiscent.*			
	Variety.....*Capsule.*			
	Dehiscence..*Loculicidal.*			
	No. of Seeds..*Many.*			
	Description of Seed..*Ovoid, with membranaceous tip. Albuminous.*			

FLORAL DIAGRAM.

CLASSIFICATION, &c.

SERIES . . . . *Phanerogams.*	Botanical Name . . . . *Erythronium Americanum.*
CLASS . . . . *Angiosperms.*	Popular Name . . . . *Dog's-tooth Violet.*
SUB-CLASS . . . . *Monocotyledons.*	Habitat . . . . *Copses.*
DIVISION . . . . *Petaloideous.*	Where found . . . . *High Park, Toronto.*
Order . . . . *Liliaceæ.*	Date of collection . . . . *May 3rd, 1894.*
Genus . . . . *Erythronium.*	
Species . . . . *Americanum.*	

DRAWINGS &C.

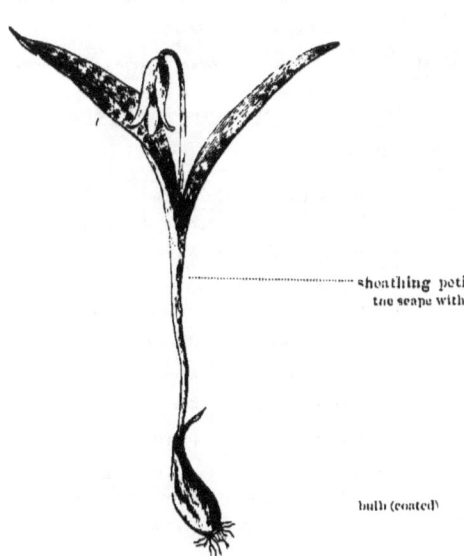
sheathing petiole, the scape within
bulb (coated)

Pistil.

cross-section of ovary.

NOTE.—*This plant sends up a scape, 5 or 6 inches high, which bears a single nodding flower. The leaves are two in number.*

| COMPOSITES. | PLANT SCHEDULE. | NO. |

ROOT.
- Origin....... *Secondary.*
- Form........ *Fibrous.*
- Colour...... *Brownish.*
- Duration.... *Perennial.*
- Position..... *Subterranean.*

STEM.
- Class........ *Dicotyledonous.*
- Attitude..... *Erect.*
- Texture..... *Herbaceous.*
- Position..... *Aerial, from a rootstock.*
- Shape....... *Cylindrical, slightly grooved.*
- Juice........ *Colourless.*
- Branching... *Usually none.*
- Height...... *About 18 inches.*
- Duration.... *Rhizome perennial; aerial stem annual.*
- Surface..... *Smooth, or nearly so.*

LEAF.
- Division.... *Simple.*
- Position.... *Radical and cauline.*
- Arrangement.. *Alternate.*
- Stipulation.. *Exstipulate.*
- Insertion... *Lower petiolate; upper sessile.*
- Outline..... *Lower spathulate; upper linear.*
- No. of leaflets, if any.. *None.*
- *Texture.... *Rather thick.*
- *Colour..... *Green both sides.*
- *Size....... *1–1½ inches long.*
- *Venation.. *Pinnately net-veined.*
- *Margin.... *Radical crenate; cauline serrate.*
- *Apex...... *Obtuse.*
- *Base...... *Lower tapering; upper clasping.*
- *Surface.... *Glabrous.*

* Applicable to leaflets if leaf is compound.

INFLORESCENCE, &c.

HEADS.
- Arrangement.. *Terminal, solitary.*
- Kind......... *Radiate.*
- Size......... *1½–2 inches across.*

RAY-FLORETS.
- Number...... *Many.*
- Colour....... *White.*
- Shape....... *Linear oblong.*
- Kind........ *Pistillate.*
- Pappus...... *Wanting.*

DISK-FLORETS.
- Number...... *Very many.*
- Colour....... *Yellow.*
- Shape....... *Tubular, slightly compressed.*
- Kind........ *Perfect.*
- Pappus...... *Wanting.*

RECEPTACLE.
- Form........ *Flattish, or slightly convex.*
- Surface...... *Naked.*

INVOLUCRE.
- Form........ *Broad and flat.*
- Rows of Scales.. *About four.*
- Form of Scales.. *Lanceolate.*
- Texture of Scales.. *With scarious margins.*
- Arrangement.. *Imbricated.*

ACHENES.
- Form........ *Nearly cylindrical.*
- Surface...... *Striate or ribbed.*
- Colour....... *Whitish or grayish.*

SEED.
- *Exalbuminous.*

If florets are all alike give particulars under heading Disk florets.

THE FLOWER.

Organ.	No.	Cohesion.	Adhesion.	Floral Diagram.
Calyx. Sepals.	5	Gamosepalous.	Superior.	
Corolla. Petals.	5	Gamopetalous.	Epigynous.	
Stamens. Filaments. Anthers.	5 5 5	Syngenesious.	Epipetalous.	
Pistil. Stigmas. Styles. Carpels. Ovary-cells.	2 1 2 1	Syncarpous.	Inferior.	

CLASSIFICATION, &c.

SERIES............ ....*Phanerogams.*
CLASS................*Angiosperms.*
SUB-CLASS............*Dicotyledons.*
DIVISION ....*Gamopetalous.*
ORDER................*Compositæ.*
GENUS.........,.....*Leucanthemum.*
SPECIES..............*Vulgare.*

Botanical Name........*Leucanthemum vulgare.*
Popular Name.........*Ox-eye Daisy.*
Habitat..............*Fields and pastures.*
Where found..........*Barrie.*
Date of collection.....*August 10th, 1863.*

DRAWINGS, &C.

LEAF SCHEDULES.

Leaf of Round-leaved Mallow.

DESCRIPTION.	DRAWINGS.
Division .... *Simple.* Position .... *Cauline.* Arrangement .... *Alternate.* Insertion .... *Petiolate.* Stipulation .... *Stipulate.* Outline .... *Orbicular.* No. of leaflets, if any ... *None.* *Texture .... *Thickish.* *Colour .... *Dark green both sides.* *Size .... *1 to 3 inches across.* *Venation .... *Palmately net-veined.* *Margin .... *Slightly lobed and crenate.* *Apex .... *Obtuse.* *Base .... *Deeply cordate.* *Surface .... *Minutely pubescent both sides.* * Applicable to leaflets if leaf is compound.	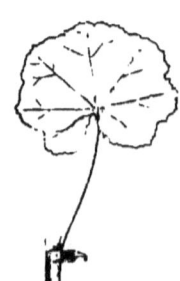

Leaf of Red Clover.

Division .... *Compound ; palmate.* Position .... *Cauline.* Arrangement .... *Alternate.* Insertion .... *Petiolate.* Stipulation .... *Stipulate ; stipules united with petiole.* Outline .... *Deltoid, or triangular.* No. of leaflets, if any ... *3.* *Texture .... *Rather thin and soft.* *Colour .... *Green, with a white spot above.* *Size .... *1 to 1½ inches long.* *Venation .... *Pinnately net-veined.* *Margin .... *Entire, or obscurely serrate.* *Apex .... *Obtuse ; emarginate.* *Base .... *Tapering.* *Surface .... *Pubescent both sides.* * Applicable to leaflets if leaf is compound.	

DESCRIPTIVE SCHEDULES.

PLANT SCHEDULE.

NO.

ROOT.	Origin		LEAF.	Division
	Form			Position
	Colour			Arrangement
	Duration			Stipulation
	Position			Insertion
STEM.	Class			Outline
	Attitude			No. of leaflets, if any
	Texture			*Texture
	Position			*Colour
	Shape			*Size
	Juice			*Venation
	Branching			*Margin
	Height			*Apex
	Duration			*Base
	Surface			*Surface

* Applicable to leaflets if leaf is compound.

INFLORESCENCE. Mode   Variety

THE FLOWER.

ORGAN.	No.	COHESION.	ADHESION.	NOTES ON FORM, ÆSTIVATION, COLOUR, ETC.
Perianth. Leaves.				
Calyx. Sepals.				
Corolla. Petals.				
Stamens. Filaments. Anthers.				
Pistil. Stigmas. Styles. Carpels. Ovary-cells.				
FRUIT. Kind Variety Dehiscence No. of Seeds Description of Seed				

FLORAL DIAGRAM.

CLASSIFICATION, &c.

SERIES..................................   Botanical Name ....
CLASS..................................   Popular Name
SUB-CLASS..........................   Habitat ....
DIVISION...............................   Where found ....
Order.................................. Date of collection
Genus...................................
Species.................................

DRAWINGS, &C.

PLANT SCHEDULE.   No._____

ROOT.	Origin		LEAF.	Division
	Form			Position
	Colour			Arrangement
	Duration			Stipulation
	Position			Insertion
STEM.	Class			Outline
	Attitude			No. of leaflets, if any
	Texture			*Texture
	Position			*Colour
	Shape			*Size
	Juice			*Venation
	Branching			*Margin
	Height			*Apex
	Duration			*Base
	Surface			*Surface

* Applicable to leaflets if leaf is compound.

INFLORESCENCE. Mode   Variety

THE FLOWER.

ORGAN.	No.	COHESION.	ADHESION.	NOTES ON FORM, ÆSTIVATION, COLOUR, ETC.

Perianth.
 Leaves.

Calyx.
 Sepals.

Corolla.
 Petals.

Stamens.
 Filaments.
 Anthers.

Pistil.
 Stigmas.
 Styles.
 Carpels.
 Ovaryes &c.

FRUIT.
  Kind
  Variety
  Dehiscence
  No. of Seeds
  Insertion of Seed

FLORAL DIAGRAM.

CLASSIFICATION, &c.

SERIES....................	Botanical Name....................
CLASS....................	Popular Name....................
SUB-CLASS....................	Habitat....................
DIVISION....................	Where found....................
Order....................	Date of collection
Genus....................	
Species....................	

DRAWINGS, &C.

PLANT SCHEDULE. No. _____

ROOT.	Origin	**LEAF.**	Division
	Form		Position
	Colour		Arrangement
	Duration		Stipulation
	Position		Insertion
STEM.	Class		Outline
	Attitude		No. of leaflets, if any
	Texture		*Texture
	Position		*Colour
	Shape		*Size
	Juice		*Venation
	Branching		*Margin
	Height		*Apex
	Duration		*Base
	Surface		*Surface

* Applicable to leaflets if leaf is compound.

INFLORESCENCE. Mode _____   Variety _____

THE FLOWER.

ORGAN.	No.	COHESION.	ADHESION.	NOTES ON FORM, ÆSTIVATION, COLOUR, ETC.
Perianth. Leaves.				
Calyx. Sepals.				
Corolla. Petals.				
Stamens. Filaments. Anthers.				
Pistil. Stigmas. Styles. Carpels. Ovary-cells.				

FRUIT.	Kind	
	Variety	
	Dehiscence	
	No. of Seeds	
	Description of Seed	

FLORAL DIAGRAM.

CLASSIFICATION, &c.

SERIES...	Botanical Name.........
CLASS............	Popular Name
SUB-CLASS...........	Habitat........
DIVISION....	Where found
Order...	Date of collection
Genus...	
Species	

DRAWINGS, &C.

PLANT SCHEDULE.   NO. _____

ROOT.	Origin		**LEAF.**	Division
	Form			Position
	Colour			Arrangement
	Duration			Stipulation
	Position			Insertion
STEM.	Class			Outline
	Attitude			No. of leaflets, if any
	Texture			*Texture
	Position			*Colour
	Shape			*Size
	Juice			*Venation
	Branching			*Margin
	Height			*Apex
	Duration			*Base
	Surface			*Surface

* Applicable to leaflets if leaf is compound.

INFLORESCENCE. Mode   Variety

THE FLOWER.

ORGAN.	No.	COHESION.	ADHESION.	NOTES ON FORM, ÆSTIVATION, COLOUR, ETC.
Perianth. Leaves.				
Calyx. Sepals.				
Corolla. Petals.				
Stamens. Filaments. Anthers.				
Pistil. Stigmas Styles Carpels Ovary cells.				
FRUIT. Kind Variety Dehiscence No. of Seeds Description of Seed				

FLORAL DIAGRAM.

CLASSIFICATION, &c.

SERIES...	Botanical Name...............................
CLASS..	Popular Name.................................
SUB-CLASS.......................................	Habitat..
DIVISION..	Where found....................................
Order...	Date of collection
Genus...	
Species..	

DRAWINGS, &C.

PLANT SCHEDULE. No...

ROOT.	Origin	**LEAF.**	Division
	Form		Position
	Colour		Arrangement
	Duration		Stipulation
	Position		Insertion
STEM.	Class		Outline
	Attitude		No. of leaflets, if any
	Texture		*Texture
	Position		*Colour
	Shape		*Size
	Juice		*Venation
	Branching		*Margin
	Height		*Apex
	Duration		*Base
	Surface		*Surface

* Applicable to leaflets if leaf is compound.

INFLORESCENCE. | Mode Variety

THE FLOWER.

ORGAN.	No.	COHESION.	ADHESION.	NOTES ON FORM, ÆSTIVATION, COLOUR, &c.
Perianth, Leaves.				
Calyx, Sepals.				
Corolla, Petals.				
Stamens, Filaments. Anthers.				
Pistil, Stigmas. Styles. Carpels. Ovary-cells.				
FRUIT. Kind Variety Dehiscence No. of Seeds Description of Seed				

FLORAL DIAGRAM.

CLASSIFICATION, &c.

SERIES......	Botanical Name
CLASS......	Popular Name
SUB-CLASS	Habitat.....
DIVISION	Where found
Order......	Date of collection
Genus.....	
Species	

DRAWINGS, &C.

PLANT SCHEDULE.   NO.....

ROOT.	Origin		LEAF.	Division
	Form			Position
	Colour			Arrangement
	Duration			Stipulation
	Position			Insertion
STEM.	Class			Outline
	Attitude			No. of leaflets, if any
	Texture			*Texture
	Position			*Colour
	Shape			*Size
	Juice			*Venation
	Branching			*Margin
	Height			*Apex
	Duration			*Base
	Surface			*Surface

* Applicable to leaflets if leaf is compound.

INFLORESCENCE. Mode   Variety

THE FLOWER.

ORGAN	N...	COHESION.	ADHESION.	NOTES ON FORM, ÆSTIVATION, COLOUR, ETC.
Perianth.				
Leaves				
Calyx.				
Sepals				
Corolla.				
Petals				
Stamens.				
Filament				
Anther				
Pistil.				
Stigma				
Style				
Carpels				
Ovary				
FRUIT.				
	Kind			
	Variety			
	Dehiscence			
	No. of Seeds			
	Description of Seed			

FLORAL DIAGRAM.

CLASSIFICATION, &c.

SERIES...	Botanical Name..............................
CLASS...	Popular Name.................................
SUB-CLASS....................................	Habitat..
DIVISION.......................................	Where found..................................
Order..	Date of collection...........................
Genus...	
Species...	

DRAWINGS, &C.

PLANT SCHEDULE.   NO. _____

ROOT.	Origin	**LEAF.**	Division _____
	Form		Position _____
	Colour		Arrangement _____
	Duration		Stipulation
	Position		Insertion
STEM.	Class		Outline
	Attitude		No. of leaflets, if any
	Texture		*Texture
	Position		*Colour
	Shape		*Size
	Juice		*Venation
	Branching		*Margin
	Height		*Apex
	Duration		*Base
	Surface		*Surface

* Applicable to leaflets if leaf is compound

INFLORESCENCE. Mode   Variety

THE FLOWER.

Organ	No.	Cohesion.	Adhesion.	Notes on Form, Æstivation, Colour, etc.
Perianth. Leaves				
Calyx. Sepals				
Corolla. Petals				
Stamens. Filament Anther				
Pistil. Stigma Styles Carpels Ovary				
FRUIT. Kind Variety Dehiscence No. of Seeds Description of Seed				

FLORAL DIAGRAM.

CLASSIFICATION, &c.

SERIES.........	Botanical Name.........
CLASS.........	Popular Name.........
SUB-CLASS.........	Habitat.........
DIVISION.........	Where found
Order.........	Date of collection
Genus.........	
Species.........	

DRAWINGS, &C.

PLANT SCHEDULE.   No._____

ROOT.	Origin		LEAF.	Division
	Form			Position
	Colour			Arrangement
	Duration			Stipulation
	Position			Insertion
STEM.	Class			Outline
	Attitude			No. of leaflets, if any
	Texture			*Texture
	Position			*Colour
	Shape			*Size
	Juice			*Venation
	Branching			*Margin
	Height			*Apex
	Duration			*Base
	Surface			*Surface

* Applicable to leaflets if leaf is compound.

INFLORESCENCE.   Mode _____   Variety _____

THE FLOWER.

ORGAN.	No.	COHESION.	ADHESION.	NOTES ON FORM, ÆSTIVATION, COLOUR, ETC.
Perianth. Leaves.				
Calyx. Sepals.				
Corolla. Petals.				
Stamens. Filaments. Anthers.				
Pistil. Stigmas. Styles. Carpels. Ovary-cells.				
FRUIT. Kind Variety Dehiscence No. of Seeds Description of Seed				

FLORAL DIAGRAM.

CLASSIFICATION, &c.

SERIES..................................	Botanical Name....................
CLASS...................................	Popular Name......................
SUB-CLASS............................	Habitat...............................
DIVISION..............................	Where found.......................
Order...................................	Date of collection................
Genus...................................	
Species.................................	

DRAWINGS, &C.

PLANT SCHEDULE.   NO. ....

ROOT.	Origin		LEAF.	Division
	Form			Position
	Colour			Arrangement
	Duration			Stipulation
	Position			Insertion
STEM.	Class			Outline
	Attitude			No. of leaflets, if any
	Texture			*Texture
	Position			*Colour
	Shape			*Size
	Juice			*Venation
	Branching			*Margin
	Height			*Apex
	Duration			*Base
	Surface			*Surface

* Applicable to leaflets if leaf is compound.

INFLORESCENCE. Mode.   Variety

THE FLOWER.

ORGAN.	No.	COHESION.	ADHESION.	NOTES ON FORM, ÆSTIVATION, COLOUR, ETC.
Perianth. *Leaves.*				
Calyx. *Sepals.*				
Corolla. *Petals.*				
Stamens. *Filaments. Anthers.*				
Pistil. *Stigmas. Styles. Carpels. Ovary-cells*				

FRUIT.	Kind
	Variety
	Dehiscence
	No. of Seeds
	Description of Seed

FLORAL DIAGRAM.

CLASSIFICATION, &c.

SERIES......   Botanical Name
CLASS...   Popular Name
SUB-CLASS ......   Habitat
DIVISION......   Where found
Order......   Date of collection
Genus ......
Species ...

DRAWINGS, &C.

PLANT SCHEDULE. NO. ____

ROOT.	Origin	**LEAF.**	Division
	Form		Position
	Colour		Arrangement
	Duration		Stipulation
	Position		Insertion
STEM.	Class		Outline
	Attitude		No. of leaflets, if any
	Texture		*Texture
	Position		*Colour
	Shape		*Size
	Juice		*Venation
	Branching		*Margin
	Height		*Apex
	Duration		*Base
	Surface		*Surface

* Applicable to leaflets if leaf is compound

INFLORESCENCE. Mode — Variety

THE FLOWER.

ORGAN.	No.	COHESION.	ADHESION.	NOTES ON FORM, ÆSTIVATION, COLOUR, ETC.
Perianth. *Leaves.*				
Calyx. *Sepals.*				
Corolla. *Petals.*				
Stamens. *Filaments.* *Anthers.*				
Pistil. *Stigmas. Styles. Carpels. Ovaries. &c.*				
FRUIT. Kind Variety Dehiscence No. of Seeds Description of Seed				

FLORAL DIAGRAM.

CLASSIFICATION. &c.

SERIES	Botanical Name
CLASS	Popular Name
SUB-CLASS	Habitat
DIVISION	Where found
Order	Date of collection
Genus	
Species	

DRAWINGS, &C.

PLANT SCHEDULE.   NO. _____

ROOT.	Origin	**LEAF.**	Division
	Form		Position
	Colour		Arrangement
	Duration		Stipulation
	Position		Insertion
STEM.	Class		Outline
	Attitude		No. of leaflets, if any
	Texture		*Texture
	Position		*Colour
	Shape		*Size
	Juice		*Venation
	Branching		*Margin
	Height		*Apex
	Duration		*Base
	Surface		*Surface
			* Applicable to leaflets if leaf is compound.
INFLORESCENCE.	Mode		Variety

THE FLOWER.

ORGAN.	No.	COHESION.	ADHESION.	NOTES ON FORM, ÆSTIVATION, COLOUR, ETC.
Perianth. Leaves.				
CALYX. Sepals.				
Corolla. Petals.				
Stamens. Filaments. Anthers.				
Pistil. Stigmas Styles Carpels Ovary-cells				
FRUIT. Kind Variety Dehiscence No. of Seeds Description of Seed				

FLORAL DIAGRAM.

CLASSIFICATION, &c.

SERIES.........	Botanical Name ......
CLASS	Popular Name
SUB-CLASS	Habitat.........
DIVISION...	Where found
Order	Date of collection
Genus.	
Species	

DRAWINGS, &C.

PLANT SCHEDULE.   NO. _____

ROOT.	Origin	**LEAF.**	Division
	Form		Position
	Colour		Arrangement
	Duration		Stipulation
	Position		Insertion
STEM.	Class		Outline
	Attitude		No. of leaflets, if any
	Texture		*Texture
	Position		*Colour
	Shape		*Size
	Juice		*Venation
	Branching		*Margin
	Height		*Apex
	Duration		*Base
	Surface		*Surface

*Applicable to leaflets if leaf is compound.

INFLORESCENCE. Mode   Variety

THE FLOWER.

ORGAN	No.	COHESION.	ADHESION.	NOTES ON FORM, ÆSTIVATION, COLOUR, ETC.
Perianth. Leaves.				
Calyx. Sepals.				
Corolla. Petals.				
Stamens. Filaments. Anthers.				
Pistil. Stigmas. Styles. Carpels. Ovary.				

FRUIT.
    Kind
    Variety
    Dehiscence
    No. of Seeds
    Description of Seed

FLORAL DIAGRAM.

CLASSIFICATION, &c.

SERIES................	Botanical Name............
CLASS.................	Popular Name.............
SUB-CLASS.......	Habitat........................
DIVISION........	Where found...............
Order...............	Date of collection
Genus..............	
Species............	

DRAWINGS, &C.

PLANT SCHEDULE.   NO.

ROOT.	Origin	**LEAF.**	Division
	Form		Position
	Colour		Arrangement
	Duration		Stipulation
	Position		Insertion
STEM.	Class		Outline
	Attitude		No. of leaflets, if any
	Texture		*Texture
	Position		*Colour
	Shape		*Size
	Juice		*Venation
	Branching		*Margin
	Height		*Apex
	Duration		*Base
	Surface		*Surface

* Applicable to leaflets if leaf is compound.

INFLORESCENCE. Mode   Variety

THE FLOWER.

ORGAN.	No.	COHESION.	ADHESION.	Notes of Form, Æstivation, Colour, etc.
Perianth. *Leaves.*				
Calyx. *Sepals.*				
Corolla. *Petals.*				
Stamens. *Filaments. Anthers.*				
Pistil. *Stigmas Styles Carpels. Ovary-cells*				

FRUIT.	Kind	
	Variety	
	Dehiscence	
	No. of Seeds	
	Description of Seed	

FLORAL DIAGRAM.

CLASSIFICATION, &c.

SERIES...............	Botanical Name................
CLASS......	Popular Name........
SUB-CLASS	Habitat......
DIVISION	Where found
Order..........	Date of collection
Genus.....	
Species	

DRAWINGS, &c.

PLANT SCHEDULE.   NO. .....

ROOT. Origin
Form
Colour
Duration
Position

STEM. Class
Attitude
Texture
Position
Shape
Juice
Branching
Height
Duration
Surface

LEAF. Division
Position
Arrangement
Stipulation
Insertion
Outline
No. of leaflets, if any
*Texture
*Colour
*Size
*Venation
*Margin
*Apex
*Base
*Surface
    * Applicable to leaflets if leaf is compound.

INFLORESCENCE. Mode   Variety

THE FLOWER.

Organs   No.   Cohesion.   Adhesion.   Notes on Form, Æstivation, Colour, etc.

Perianth.
  Leaves.

Calyx.
  Sepals.

Corolla.
  Petals.

Stamens.
  Filaments.
  Anthers.

Pistil.
  Stigmas.
  Styles.
  Carpels.
  Ovary cells.

FRUIT. Kind
Variety
Dehiscence
No. of Seeds
Description of Seed

FLORAL DIAGRAM

CLASSIFICATION, &c.

SERIES	Botanical Name
CLASS	Popular Name
SUB-CLASS	Habitat
DIVISION	Where found
Order	Date of collection
Genus	
Species	

DRAWINGS, &C.

PLANT SCHEDULE.   NO. ....

ROOT.	Origin		LEAF.	Division
	Form			Position
	Colour			Arrangement
	Duration			Stipulation
	Position			Insertion
STEM.	Class			Outline
	Attitude			No. of leaflets, if any
	Texture			*Texture
	Position			*Colour
	Shape			*Size
	Juice			*Venation
	Branching			*Margin
	Height			*Apex
	Duration			*Base
	Surface			*Surface

* Applicable to leaflets if leaf is compound.

INFLORESCENCE. Mode   Variety

THE FLOWER.

ORGAN	No.	Cohesion.	Adhesion.	Notes on Form, Æstivation, Colour, etc.

Perianth.
 Leaves.

Calyx.
 Sepals.

Corolla.
 Petals.

Stamens.
 Filaments.
 Anthers.

Pistil.
 Stigma
 Styles.
 Carpels.
 Ovary.

FRUIT.
  Kind
  Variety
  Dehiscence
  No. of Seeds
  Description of Seed

FLORAL DIAGRAM

CLASSIFICATION, &c.

SERIES	Botanical Name ....
CLASS	Popular Name ....
SUB-CLASS	Habitat ....
DIVISION	Where found ....
Order	Date of collection ....
Genus	
Species	

DRAWINGS, &C.

PLANT SCHEDULE.   NO.

ROOT.	Origin		**LEAF.**	Division
	Form			Position
	Colour			Arrangement
	Duration			Stipulation
	Position			Insertion
STEM.	Class			Outline
	Attitude			No. of leaflets, if any
	Texture			*Texture
	Position			*Colour
	Shape			*Size
	Juice			*Venation
	Branching			*Margin
	Height			*Apex
	Duration			*Base
	Surface			*Surface

* Applicable to leaflets if leaf is compound.

INFLORESCENCE. Mode ——— Variety ———

THE FLOWER.

ORGAN	NO.	COHESION.	ADHESION.	NOTES ON FORM, ÆSTIVATION, COLOUR, ETC.
Perianth. Leaves.				
Calyx. Sepals.				
Corolla. Petals.				
Stamens. Filaments. Anthers.				
Pistil. Stigmas. Styles. Carpels. Ovary-cells.				
FRUIT.	Kind			
	Variety			
	Dehiscence			
	No. of Seeds			
	Description of Seed			

FLORAL DIAGRAM.

CLASSIFICATION, &c.

SERIES...............	Botanical Name
CLASS............	Popular Name....
SUB-CLASS..	Habitat............
DIVISION........	Where found
Order.........	Date of collection
Genus............	
Species.........	

DRAWINGS, &C.

PLANT SCHEDULE.   No. ....

ROOT.	Origin		LEAF.	Division
	Form			Position
	Colour			Arrangement
	Duration			Stipulation
	Position			Insertion
STEM.	Class			Outline
	Attitude			No. of leaflets, if any
	Texture			*Texture
	Position			*Colour
	Shape			*Size
	Juice			*Venation
	Branching			*Margin
	Height			*Apex
	Duration			*Base
	Surface			*Surface

* Applicable to leaflets if leaf is compound.

INFLORESCENCE. | Mode ......   Variety ......

THE FLOWER.

ORGAN.	No.	COHESION.	ADHESION.	NOTES ON FORM, ÆSTIVATION, COLOUR, ETC.
Perianth. *Leaves.*				
Calyx. *Sepals.*				
Corolla. *Petals.*				
Stamens. *Filaments.* *Anthers.*				
Pistil. *Stigmas* *Styles* *Carpels.* *Ovary-cells.*				

FRUIT.
  Kind
  Variety
  Dehiscence
  No. of Seeds
  Description of Seed

FLORAL DIAGRAM.

CLASSIFICATION, &c.

SERIES	Botanical Name ....
CLASS	Popular Name
SUB-CLASS	Habitat ....
DIVISION	Where found ....
Order	Date of collection
Genus	
Species	

DRAWINGS, &C.

PLANT SCHEDULE.   NO. _____

ROOT. Origin
Form
Colour
Duration
Position

STEM. Class
Attitude
Texture
Position
Shape
Juice
Branching
Height
Duration
Surface

LEAF. Division
Position
Arrangement
Stipulation
Insertion
Outline
No. of leaflets, if any
*Texture
*Colour
*Size
*Venation
*Margin
*Apex
*Base
*Surface
* Applicable to leaflets if leaf is compound.

INFLORESCENCE. Mode   Variety

THE FLOWER.

ORGAN	No.	COHESION.	ADHESION.	NOTES ON FORM, ÆSTIVATION, COLOUR, ETC.
Perianth. Leaves.				
Calyx. Sepals.				
Corolla. Petals.				
Stamens. Filaments. Anthers.				
Pistil. Stigmas. Styles. Carpels. Ovaries.				

FRUIT. Kind
Variety
Dehiscence
No. of Seeds
Description of Seed

FLORAL DIAGRAM.

CLASSIFICATION, &c.

SERIES		Botanical Name
CLASS		Popular Name
SUB-CLASS		Habitat
DIVISION		Where found
Order		Date of collection
Genus		
Species		

DRAWINGS, &C.

PLANT SCHEDULE.   NO..........

ROOT.	Origin	**LEAF.**	Division
	Form.		Position
	Colour		Arrangement
	Duration		Stipulation
	Position		Insertion
STEM.	Class		Outline
	Attitude		No. of leaflets, if any
	Texture		*Texture
	Position		*Colour
	Shape		*Size
	Juice		*Venation
	Branching		*Margin
	Height		*Apex
	Duration		*Base
	Surface		*Surface

* Applicable to leaflets if leaf is compound.

INFLORESCENCE.   Mode .... Variety ....

THE FLOWER.

ORGAN.	No.	COHESION.	ADHESION.	NOTES ON FORM, ÆSTIVATION, COLOUR, ETC.
Perianth. Leaves.				
Calyx. Sepals.				
Corolla. Petals.				
Stamens. Filaments. Anthers.				
Pistil. Stigmas. Styles. Carpels. Ovary-cells.				
FRUIT. Kind Variety Dehiscence No. of Seeds Description of Seed				

FLORAL DIAGRAM.

CLASSIFICATION, &c.

SERIES........................	Botanical Name........................
CLASS........................	Popular Name........................
SUB-CLASS................	Habitat........................
DIVISION.	Where found
Order....	Date of collection
Genus	
Species	

DRAWINGS, &C.

PLANT SCHEDULE.   NO.............

ROOT.	Origin	**LEAF.**	Division
	Form		Position
	Colour		Arrangement
	Duration		Stipulation
	Position		Insertion
STEM.	Class		Outline
	Attitude		No. of leaflets, if any
	Texture		*Texture
	Position		*Colour
	Shape		*Size
	Juice		*Venation
	Branching		*Margin
	Height		*Apex
	Duration		*Base
	Surface		*Surface

* Applicable to leaflets if leaf is compound.

INFLORESCENCE. Mode   Variety

THE FLOWER.

ORGAN.	No.	COHESION.	ADHESION.	NOTES ON FORM, ÆSTIVATION, COLOUR, ETC.
Perianth, Leaves.				
Calyx, Sepals.				
Corolla, Petals.				
Stamens, Filament, Anther.				
Pistil, Stigma, Style, Carpels, Ovaries.				
FRUIT. Kind, Variety, Dehiscence, No. of Seeds, Description of Seed				

FLORAL DIAGRAM.

CLASSIFICATION, &c.

SERIES...............	Botanical Name...............
CLASS...............	Popular Name...............
SUB-CLASS...............	Habitat...............
DIVISION...............	Where found...............
Order...............	Date of collection...............
Genus...............	
Species...............	

DRAWINGS, &C.

| COMPOSITES. | PLANT SCHEDULE. | NO. |

ROOT.	Origin	LEAF.	Division
	Form		Position
	Colour		Arrangement
	Duration		Stipulation
	Position		Insertion
STEM.	Class		Outline
	Attitude		No. of leaflets, if any
	Texture		*Texture
	Position		*Colour
	Shape		*Size
	Juice		*Venation
	Branching		*Margin
	Height		*Apex
	Duration		*Base
	Surface		*Surface

* Applicable to leaflets if leaf is compound.

INFLORESCENCE, &c.

HEADS.		RECEPTACLE.
Arrangement		Form
Kind		Surface
Size		INVOLUCRE.
*RAY-FLORETS.		Form
Number		Rows of scales
Colour		Form of scales
Shape		Texture of scales
Kind		Arrangement
Pappus		ACHENES.
*DISK-FLORETS.		Form
Number		Surface
Colour		Colour
Shape		SEED.
Kind		
Pappus		

* If florets are all alike give particulars under heading Disk-florets.

THE FLOWER.

Organ.	No.	Cohesion.	Adhesion.	Floral Diagram.
Calyx. Sepals.				
Corolla. Petals.				
Stamens. Filaments. Anthers.				
Pistil. Stigmas. Styles. Carpels. Ovary-cells.				

CLASSIFICATION, &c.

SERIES............	Botanical Name............
CLASS............	Popular Name............
SUB-CLASS............	Habitat............
DIVISION............	Where found............
Order............	Date of collection
Genus............	
Species............	

DRAWINGS, &C.

COMPOSITES. PLANT SCHEDULE. NO._____

ROOT.
- Origin
- Form
- Colour
- Duration
- Position

STEM.
- Class
- Attitude
- Texture
- Position
- Shape
- Juice
- Branching
- Height
- Duration
- Surface

LEAF.
- Division
- Position
- Arrangement
- Stipulation
- Insertion
- Outline
- No. of leaflets, if any
- *Texture
- *Colour
- *Size
- *Venation
- *Margin
- *Apex
- *Base
- *Surface

* Applicable to leaflets if leaf is compound.

INFLORESCENCE, &c.

HEADS.
- Arrangement
- Kind
- Size

*RAY-FLORETS.
- Number
- Colour
- Shape
- Kind
- Pappus

*DISK-FLORETS.
- Number
- Colour
- Shape
- Kind
- Pappus

RECEPTACLE.
- Form
- Surface

INVOLUCRE.
- Form
- Rows of scales
- Form of scales
- Texture of scales
- Arrangement

ACHENES.
- Form
- Surface
- Colour

SEED.

* If florets are all alike give particulars under heading Disk-florets.

THE FLOWER.

Organ.	No.	Cohesion.	Adhesion.	Floral Diagram.
Calyx. *Sepals.*				
Corolla. *Petals.*				
Stamens. *Filaments. Anthers.*				
Pistil. *Stigmas. Styles. Carpels. Ovary-cells.*				

CLASSIFICATION, &c.

SERIES ....
CLASS ....
SUB-CLASS ....
DIVISION ....
Order ....
Genus ....
Species ....

Botanical Name ....
Popular Name ....
Habitat ....
Where found ....
Date of collection ....

DRAWINGS, &c.

| COMPOSITES. | PLANT SCHEDULE. | NO. |

ROOT.	Origin	**LEAF.**	Division
	Form		Position
	Colour		Arrangement
	Duration		Stipulation
	Position		Insertion
STEM.	Class		Outline
	Attitude		No. of leaflets, if any
	Texture		*Texture
	Position		*Colour
	Shape		*Size
	Juice		*Venation
	Branching		*Margin
	Height		*Apex
	Duration		*Base
	Surface		*Surface
			* Applicable to leaflets if leaf is compound.

INFLORESCENCE, &c.

HEADS.		**RECEPTACLE.**	
	Arrangement		Form
	Kind		Surface
	Size	**INVOLUCRE.**	
RAY-FLORETS.			Form
	Number		Rows of scales
	Colour		Form of scales
	Shape		Texture of scales
	Kind		Arrangement
	Pappus	**ACHENES.**	
DISK-FLORETS.			Form
	Number		Surface
	Colour		Colour
	Shape	**SEED.**	
	Kind		
	Pappus		If florets are all alike give particulars under heading Disk-florets.

THE FLOWER.

Organ.	No.	Cohesion.	Adhesion.	Floral Diagram.
Calyx. *Sepals.*				
Corolla. *Petals.*				◯
Stamens. *Filaments.* *Anthers.*				
Pistil. *Stigmas.* *Styles.* *Carpels.* *Ovary-cells.*				

CLASSIFICATION, &c.

SERIES .... Botanical Name ....

CLASS .... Popular Name ....

SUB-CLASS .... Habitat ....

DIVISION .... Where found ....

Order .... Date of collection ....

Genus ....

Species ....

DRAWINGS, &C.

COMPOSITES.		PLANT SCHEDULE.		NO.
ROOT.	Origin		**LEAF.**	Division
	Form			Position
	Colour			Arrangement
	Duration			Stipulation
	Position			Insertion
STEM.	Class			Outline
	Attitude			No. of leaflets, if any
	Texture			*Texture
	Position			*Colour
	Shape			*Size
	Juice			*Venation
	Branching			*Margin
	Height			*Apex
	Duration			*Base
	Surface			*Surface
				* Applicable to leaflets if leaf is compound.

INFLORESCENCE, &c.

HEADS.		**RECEPTACLE.**
Arrangement		Form
Kind		Surface
Size		**INVOLUCRE.**
*RAY-FLORETS.		Form
Number		Rows of scales
Colour		Form of scales
Shape		Texture of scales
Kind		Arrangement
Pappus		**ACHENES.**
*DISK-FLORETS.		Form
Number		Surface
Colour		Colour
Shape		**SEED.**
Kind		
Pappus		

* If florets are all alike give particulars under heading Disk-florets.

THE FLOWER.

Organ.	No.	Cohesion.	Adhesion.	Floral Diagram.
Calyx. *Sepals.*				
Corolla. *Petals.*				
Stamens. *Filaments.* *Anthers.*				
Pistil. *Stigmas.* *Styles.* *Carpels.* *Ovary-cells.*				

CLASSIFICATION, &c.

SERIES..   Botanical Name..

CLASS...........   Popular Name..

SUB-CLASS   Habitat...

DIVISION...................   Where found..

Order..   Date of collection

Genus..........

Species..

DRAWINGS, &C.

| COMPOSITES. | PLANT SCHEDULE. | NO. _____ |

ROOT.	Origin —		LEAF.	Division
	Form			Position
	Colour			Arrangement
	Duration			Stipulation
	Position			Insertion
STEM.	Class			Outline
	Attitude			No. of leaflets, if any
	Texture			*Texture
	Position			*Colour
	Shape			*Size
	Juice			*Venation
	Branching			*Margin
	Height			*Apex
	Duration			*Base
	Surface			*Surface

* Applicable to leaflets if leaf is compound.

INFLORESCENCE, &c.

HEADS.		RECEPTACLE.	
Arrangement		Form	
Kind		Surface	
Size		INVOLUCRE.	
*RAY-FLORETS.		Form	
Number		Rows of scales	
Colour		Form of scales	
Shape		Texture of scales	
Kind		Arrangement	
Pappus		ACHENES.	
*DISK-FLORETS.		Form	
Number		Surface	
Colour		Colour	
Shape		SEED.	
Kind			
Pappus			

* If florets are all alike give particulars under heading Disk-florets.

THE FLOWER.

ORGAN.	No.	COHESION.	ADHESION.	FLORAL DIAGRAM.
Calyx. *Sepals.*				
Corolla. *Petals.*				
Stamens. *Filaments.* *Anthers.*				
Pistil. *Stigmas.* *Styles.* *Carpels.* *Ovary-cells.*				

CLASSIFICATION, &c.

SERIES	Botanical Name ....
CLASS	Popular Name ....
SUB-CLASS	Habitat ....
DIVISION	Where found ....
Order	Date of collection ....
Genus	
Species	

DRAWINGS, &C.

| COMPOSITES. | PLANT SCHEDULE. | NO. |

ROOT.	Origin	**LEAF.**	Division
	Form		Position
	Colour		Arrangement
	Duration		Stipulation
	Position		Insertion
STEM.	Class		Outline
	Attitude		No. of leaflets, if any
	Texture		*Texture
	Position		*Colour
	Shape		*Size
	Juice		*Venation
	Branching		*Margin
	Height		*Apex
	Duration		*Base
	Surface		*Surface

* Applicable to leaflets if leaf is compound.

INFLORESCENCE, &c.

HEADS.		**RECEPTACLE.**	
	Arrangement		Form
	Kind		Surface
	Size	**INVOLUCRE.**	
RAY-FLORETS.			Form
	Number		Rows of scales
	Colour		Form of scales
	Shape		Texture of scales
	Kind		Arrangement
	Pappus	**ACHENES.**	
DISK-FLORETS.			Form
	Number		Surface
	Colour		Colour
	Shape	**SEED.**	
	Kind		
	Pappus		

* If florets are all alike give particulars under heading Disk-florets.

THE FLOWER.

Organ.	No.	Cohesion.	Adhesion.	Floral Diagram.
Calyx. *Sepals.*				
Corolla. *Petals.*				
Stamens. *Filaments.* *Anthers.*				
Pistil. *Stigmas.* *Styles.* *Carpels.* *Ovary-cells.*				

CLASSIFICATION, &c.

Series ....
Class ....
Sub-Class ....
Division ....
Order ....
Genus ....
Species ....

Botanical Name ....
Popular Name ....
Habitat ....
Where found ....
Date of collection ....

DRAWINGS, &c.

LEAF SCHEDULES.

Leaf of

Description.	Drawings.
Division	
Position	
Arrangement	
Insertion	
Stipulation	
Outline	
No. of leaflets, if any	
*Texture	
*Colour	
*Size	
*Venation	
*Margin	
*Apex	
*Base	
*Surface	

* Applicable to leaflets if leaf is compound.

Leaf of

Description.	Drawings.
Division	
Position	
Arrangement	
Insertion	
Stipulation	
Outline	
No. of leaflets, if any	
*Texture	
*Colour	
*Size	
*Venation	
*Margin	
*Apex	
*Base	
*Surface —	

* Applicable to leaflets if leaf is compound.

LEAF SCHEDULES.

Leaf of ......

Description.	Drawings.
Division	
Position	
Arrangement	
Insertion	
Stipulation	
Outline	
No. of leaflets, if any	
*Texture	
*Colour	
*Size	
*Venation	
*Margin	
*Apex	
*Base	
*Surface	
* Applicable to leaflets if leaf is compound.	

Leaf of ......

Division	
Position	
Arrangement	
Insertion	
Stipulation	
Outline	
No. of leaflets, if any	
*Texture	
*Colour ...	
*Size	
*Venation	
*Margin	
*Apex	
*Base	
*Surface	
* Applicable to leaflets if leaf is compound.	

LEAF SCHEDULES.

LEAF OF _____

DESCRIPTION.	DRAWINGS.
Division	
Position	
Arrangement	
Insertion	
Stipulation	
Outline _____	
No. of leaflets, if any	
*Texture	
*Colour	
*Size —	
*Venation	
*Margin	
*Apex	
*Base	
*Surface	

* Applicable to leaflets if leaf is compound.

LEAF OF _____

DESCRIPTION.	DRAWINGS.
Division	
Position	
Arrangement	
Insertion	
Stipulation	
Outline	
No. of leaflets, if any	
*Texture	
*Colour	
*Size —	
*Venation	
*Margin	
*Apex	
*Base	
*Surface	

* Applicable to leaflets if leaf is compound.

LEAF SCHEDULES.

Leaf of ....

Description.	Drawings.
Division	
Position	
Arrangement	
Insertion	
Stipulation	
Outline	
No. of leaflets, if any	
*Texture	
*Colour	
*Size	
*Venation	
*Margin	
*Apex	
*Base	
*Surface	

* Applicable to leaflets if leaf is compound.

Leaf of ....

Description	Drawings
Division	
Position	
Arrangement	
Insertion	
Stipulation	
Outline	
No. of leaflets, if any	
*Texture	
*Colour	
*Size	
*Venation	
*Margin	
*Apex	
*Base	
*Surface	

* Applicable to leaflets if leaf is compound.

LEAF SCHEDULES.

Leaf of

Description.	Drawings.
Division	
Position	
Arrangement	
Insertion	
Stipulation	
Outline	
No. of leaflets, if any	
*Texture	
*Colour	
*Size	
*Venation	
*Margin	
*Apex	
*Base	
*Surface	

* Applicable to leaflets if leaf is compound.

Leaf of

Description	Drawings
Division	
Position	
Arrangement	
Insertion	
Stipulation	
Outline	
No. of leaflets, if any	
*Texture	
*Colour	
*Size	
*Venation	
*Margin	
*Apex	
*Base	
*Surface	

* Applicable to leaflets if leaf is compound.

LEAF SCHEDULES.

Leaf of ....

Description.	Drawings.
Division	
Position	
Arrangement	
Insertion	
Stipulation	
Outline	
No. of leaflets, if any	
*Texture	
*Colour	
*Size	
*Venation	
*Margin	
*Apex	
*Base	
*Surface	

* Applicable to leaflets if leaf is compound.

Leaf of ....

Division	
Position	
Arrangement	
Insertion	
Stipulation	
Outline	
No. of leaflets, if any	
*Texture	
*Colour	
*Size	
*Venation	
*Margin	
*Apex	
*Base	
*Surface	

* Applicable to leaflets if leaf is compound.

LEAF SCHEDULES.

LEAF OF ....

DESCRIPTION.	DRAWINGS.
Division	
Position	
Arrangement	
Insertion	
Stipulation	
Outline	
No. of leaflets, if any	
*Texture	
*Colour	
*Size	
*Venation	
*Margin	
*Apex	
*Base	
*Surface	

* Applicable to leaflets if leaf is compound.

LEAF OF ....

DESCRIPTION.	DRAWINGS.
Division	
Position	
Arrangement	
Insertion	
Stipulation	
Outline	
No. of leaflets, if any	
*Texture	
*Colour	
*Size	
*Venation	
*Margin	
*Apex	
*Base	
*Surface	

* Applicable to leaflets if leaf is compound.

LEAF SCHEDULES.

LEAF OF ....

DESCRIPTION.	DRAWINGS.
Division	
Position	
Arrangement	
Insertion	
Stipulation	
Outline	
No. of leaflets, if any	
*Texture	
*Colour	
*Size	
*Venation	
*Margin	
*Apex	
*Base	
*Surface	
* Applicable to leaflets if leaf is compound.	

LEAF OF ....

DESCRIPTION.	DRAWINGS.
Division	
Position	
Arrangement	
Insertion	
Stipulation	
Outline	
No. of leaflets, if any	
*Texture	
*Colour	
*Size	
*Venation	
*Margin	
*Apex	
*Base	
*Surface	
* Applicable to leaflets if leaf is compound.	

LEAF SCHEDULES.

Leaf of _____

DESCRIPTION.	DRAWINGS.
Division	
Position	
Arrangement	
Insertion	
Stipulation	
Outline	
No. of leaflets, if any	
*Texture	
*Colour	
*Size	
*Venation	
*Margin	
*Apex	
*Base	
*Surface	

* Applicable to leaflets if leaf is compound

Leaf of _____

DESCRIPTION.	DRAWINGS.
Division	
Position	
Arrangement	
Insertion	
Stipulation	
Outline	
No. of leaflets, if any	
*Texture	
*Colour	
*Size	
*Venation	
*Margin	
*Apex	
*Base	
*Surface	

* Applicable to leaflets if leaf is compound

LEAF SCHEDULES.

LEAF OF ....

Description.	Drawings.
Division	
Position	
Arrangement	
Insertion	
Stipulation	
Outline	
No. of leaflets, if any	
*Texture	
*Colour	
*Size	
*Venation	
*Margin	
*Apex	
*Base	
*Surface	

*Applicable to leaflets if leaf is compound.

LEAF OF ....

Description.	Drawings.
Division	
Position	
Arrangement	
Insertion	
Stipulation	
Outline	
No. of leaflets, if any	
*Texture	
*Colour	
*Size	
*Venation	
*Margin	
*Apex	
*Base	
*Surface	

*Applicable to leaflets if leaf is compound.

LEAF SCHEDULES.

MAKE DRAWINGS OF LEAVES ANSWERING TO THE FOLLOWING DESCRIPTIONS.
(The teacher will dictate the descriptions.)

DESCRIPTION.	DRAWINGS.
Division	
Position	
Arrangement	
Insertion	
Stipulation	
Outline	
No. of leaflets, if any	
*Texture	
*Colour	
*Size	
*Venation	
*Margin	
*Apex	
*Base	
*Surface	

* Applicable to leaflets if leaf is compound.

Division	
Position	
Arrangement	
Insertion	
Stipulation	
Outline	
No. of leaflets, if any	
*Texture	
*Colour	
*Size	
*Venation	
*Margin	
*Apex	
*Base	
*Surface	

* Applicable to leaflets if leaf is compound.

LEAF SCHEDULES.

MAKE DRAWINGS OF LEAVES ANSWERING TO THE FOLLOWING DESCRIPTIONS.

(The teacher will dictate the descriptions.)

DESCRIPTION.	DRAWINGS.
Division	
Position	
Arrangement	
Insertion	
Stipulation	
Outline	
No. of leaflets, if any	
*Texture	
*Colour	
*Size	
*Venation	
*Margin	
*Apex	
*Base	
*Surface	
* Applicable to leaflets if leaf is compound.	
Division	
Position	
Arrangement	
Insertion	
Stipulation	
Outline	
No. of leaflets, if any	
*Texture	
*Colour	
*Size	
*Venation	
*Margin	
*Apex	
*Base	
*Surface	
* Applicable to leaflets if leaf is compound.	

LEAF SCHEDULES.

MAKE DRAWINGS OF LEAVES ANSWERING TO THE FOLLOWING DESCRIPTIONS.

(The teacher will dictate the descriptions.)

DESCRIPTION.	DRAWINGS.
Division	
Position	
Arrangement	
Insertion	
Stipulation	
Outline	
No. of leaflets, if any	
*Texture	
*Colour	
*Size	
*Venation	
*Margin	
*Apex	
*Base	
*Surface	

* Applicable to leaflets if leaf is compound.

Division	
Position	
Arrangement	
Insertion	
Stipulation	
Outline	
No. of leaflets, if any	
*Texture	
*Colour	
*Size	
*Venation	
*Margin	
*Apex	
*Base	
*Surface	

* Applicable to leaflets if leaf is compound.

LEAF SCHEDULES.

MAKE DRAWINGS OF LEAVES ANSWERING TO THE FOLLOWING DESCRIPTIONS.

(The teacher will dictate the descriptions.)

DESCRIPTION.	DRAWINGS.
Division	
Position	
Arrangement	
Insertion	
Stipulation	
Outline	
No. of leaflets, if any	
*Texture	
*Colour	
*Size	
*Venation	
*Margin	
*Apex	
*Base	
*Surface	
* Applicable to leaflets if leaf is compound.	
Division	
Position	
Arrangement	
Insertion	
Stipulation	
Outline	
No. of leaflets, if any	
*Texture	
*Colour	
*Size	
*Venation	
*Margin	
*Apex	
*Base	
*Surface	
* Applicable to leaflets if leaf is compound.	

LEAF SCHEDULES.

Make Drawings of Leaves Answering to the Following Descriptions.
(The teacher will dictate the descriptions.)

Description.	Drawings.
Division	
Position	
Arrangement	
Insertion	
Stipulation	
Outline	
No. of leaflets, if any	
*Texture	
*Colour	
*Size	
*Venation	
*Margin	
*Apex	
*Base	
*Surface	

* Applicable to leaflets if leaf is compound.

Division	
Position	
Arrangement	
Insertion	
Stipulation	
Outline	
No. of leaflets, if any	
*Texture	
*Colour	
*Size	
*Venation	
*Margin	
*Apex	
*Base	
*Surface	

* Applicable to leaflets if leaf is compound.

LEAF SCHEDULES.

MAKE DRAWINGS OF LEAVES ANSWERING TO THE FOLLOWING DESCRIPTIONS.

(The teacher will dictate the descriptions.)

DESCRIPTION.	DRAWINGS.
Division	
Position	
Arrangement	
Insertion	
Stipulation	
Outline	
No. of leaflets, if any	
*Texture	
*Colour ..	
*Size	
*Venation	
*Margin	
*Apex	
*Base	
*Surface	
* Applicable to leaflets if leaf is compound.	
Division	
Position	
Arrangement	
Insertion	
Stipulation	
Outline	
No. of leaflets, if any	
*Texture	
*Colour	
*Size	
*Venation	
*Margin	
*Apex	
*Base	
*Surface	
* Applicable to leaflets if leaf is compound.	

LEAF SCHEDULES.

MAKE DRAWINGS OF LEAVES ANSWERING TO THE FOLLOWING DESCRIPTIONS.

(The teacher will dictate the descriptions.)

DESCRIPTION.	DRAWINGS.
Division	
Position	
Arrangement	
Insertion	
Stipulation	
Outline	
No. of leaflets, if any	
*Texture	
*Colour	
*Size	
*Venation	
*Margin	
*Apex	
*Base	
*Surface	

* Applicable to leaflets if leaf is compound.

DESCRIPTION.	DRAWINGS.
Division	
Position	
Arrangement	
Insertion	
Stipulation	
Outline	
No. of leaflets, if any	
*Texture	
*Colour	
*Size	
*Venation	
*Margin	
*Apex	
*Base	
*Surface	

* Applicable to leaflets if leaf is compound.

LEAF SCHEDULES.

MAKE DRAWINGS OF LEAVES ANSWERING TO THE FOLLOWING DESCRIPTIONS.

(The teacher will dictate the descriptions.)

DESCRIPTION.	DRAWINGS.
Division	
Position	
Arrangement	
Insertion	
Stipulation	
Outline	
No. of leaflets, if any	
*Texture	
*Colour	
*Size	
*Venation	
*Margin	
*Apex	
*Base	
*Surface	
* Applicable to leaflets if leaf is compound.	
Division	
Position	
Arrangement	
Insertion	
Stipulation	
Outline	
No. of leaflets, if any	
*Texture	
*Colour	
*Size	
*Venation	
*Margin	
*Apex	
*Base	
*Surface	
* Applicable to leaflets if leaf is compound.	

LEAF SCHEDULES.

MAKE DRAWINGS OF LEAVES ANSWERING TO THE FOLLOWING DESCRIPTIONS.

(The teacher will dictate the descriptions.)

DESCRIPTION.	DRAWINGS.
Division	
Position	
Arrangement	
Insertion	
Stipulation	
Outline	
No. of leaflets, if any	
*Texture	
*Colour	
*Size	
*Venation	
*Margin	
*Apex	
*Base	
*Surface	

* Applicable to leaflets if leaf is compound.

DESCRIPTION.	DRAWINGS.
Division	
Position	
Arrangement	
Insertion	
Stipulation	
Outline	
No. of leaflets, if any	
*Texture	
*Colour	
*Size	
*Venation	
*Margin	
*Apex	
*Base	
*Surface	

* Applicable to leaflets if leaf is compound.

LEAF SCHEDULES.

MAKE DRAWINGS OF LEAVES ANSWERING TO THE FOLLOWING DESCRIPTIONS.

(The teacher will dictate the descriptions.)

DESCRIPTION.	DRAWINGS.
Division	
Position	
Arrangement	
Insertion	
Stipulation	
Outline	
No. of leaflets, if any	
*Texture	
*Colour	
*Size	
*Venation	
*Margin	
*Apex	
*Base	
*Surface	
* Applicable to leaflets if leaf is compound.	
Division	
Position	
Arrangement	
Insertion	
Stipulation	
Outline	
No. of leaflets, if any	
*Texture	
*Colour	
*Size	
*Venation	
*Margin	
*Apex	
*Base	
*Surface	
* Applicable to leaflets if leaf is compound.	

FLOWER SCHEDULES.

Flower of _____

Organ.	No.	Cohesion.	Adhesion.	Notes on Form, Æstivation, Colour, etc.
Perianth. Leaves.				
Calyx. Sepals.				
Corolla. Petals.				
Stamens. Filaments. Anthers.				
Pistil. Stigmas. Styles. Carpels. Ovary cells.				
Fruit. Kind Variety Dehiscence No. of Seeds Description of Seed				

Floral Diagram.

Flower of _____

Organ.	No.	Cohesion.	Adhesion.	Notes on Form, Æstivation, Colour, etc.
Perianth. Leaves.				
Calyx. Sepals.				
Corolla. Petals.				
Stamens. Filaments. Anthers.				
Pistil. Stigmas. Styles. Carpels. Ovary cells.				
Fruit. Kind Variety Dehiscence No. of Seeds Description of Seed				

Floral Diagram.

FLOWER SCHEDULES.

Flower of ....

ORGAN.	No.	COHESION.	ADHESION.	NOTES ON FORM, ÆSTIVATION, COLOUR, ETC.
Perianth. *Leaves.*				
Calyx. *Sepals.*				
Corolla. *Petals.*				
Stamens. *Filaments. Anthers.*				
Pistil. *Stigmas. Styles. Carpels. Ovary-cells.*				

FRUIT.
Kind ....
Variety ....
Dehiscence ....
No. of Seeds ....
Description of Seed ....

FLORAL DIAGRAM.

Flower of ....

ORGAN.	No.	COHESION.	ADHESION.	NOTES ON FORM, ÆSTIVATION, COLOUR, ETC.
Perianth. *Leaves.*				
Calyx. *Sepals.*				
Corolla. *Petals.*				
Stamens. *Filaments. Anthers.*				
Pistil. *Stigmas. Styles. Carpels. Ovary-cells.*				

FRUIT.
Kind ....
Variety ....
Dehiscence ....
No. of Seeds ....
Description of Seed ....

FLORAL DIAGRAM.

FLOWER SCHEDULES.

FLOWER OF ____

ORGAN.	No.	COHESION.	ADHESION.	NOTES ON FORM, ÆSTIVATION, COLOUR, ETC.
Perianth. *Leaves.*				
Calyx. *Sepals.*				
Corolla. *Petals.*				
Stamens. *Filaments. Anthers.*				
Pistil. *Stigmas. Styles. Carpels. Ovary-cells.*				

FRUIT.
- Kind
- Variety
- Dehiscence
- No. of Seeds
- Description of Seed

FLORAL DIAGRAM.

FLOWER OF ____

ORGAN.	No.	COHESION.	ADHESION.	NOTES ON FORM, ÆSTIVATION, COLOUR, ETC.
Perianth. *Leaves.*				
Calyx. *Sepals.*				
Corolla. *Petals.*				
Stamens. *Filaments. Anthers.*				
Pistil. *Stigmas. Styles. Carpels. Ovary-cells.*				

FRUIT.
- Kind
- Variety
- Dehiscence
- No. of Seeds
- Description of Seed

FLORAL DIAGRAM.

FLOWER SCHEDULES.

Flower of ....

Organ.	No.	Cohesion.	Adhesion.	Notes on Form, Æstivation, Colour, etc.
Perianth. *Leaves.*				
Calyx. *Sepals.*				
Corolla. *Petals.*				
Stamens. *Filaments. Anthers.*				
Pistil. *Stigmas. Styles. Carpels. Ovary-cells.*				
Fruit.	Kind .... Variety .... Dehiscence .... No. of Seeds .... Description of Seed			

Floral Diagram.

Flower of ....

Organ.	No.	Cohesion.	Adhesion.	Notes on Form, Æstivation, Colour, etc.
Perianth. *Leaves.*				
Calyx. *Sepals.*				
Corolla. *Petals.*				
Stamens. *Filaments. Anthers.*				
Pistil. *Stigmas. Styles. Carpels. Ovary-cells.*				
Fruit.	Kind .... Variety .... Dehiscence .... No. of Seeds .... Description of Seed			

Floral Diagram.

FLOWER SCHEDULES.

Flower of _____

Organ.	No.	Cohesion.	Adhesion.	Notes on Form, Æstivation, Colour, etc.
Perianth. *Leaves.*				
Calyx. *Sepals.*				
Corolla. *Petals.*				
Stamens. *Filaments. Anthers.*				
Pistil. *Stigmas. Styles. Carpels. Ovary-cells.*				
Fruit. Kind Variety Dehiscence No. of Seeds Description of Seed				

Floral Diagram.

Flower of _____

Organ.	No.	Cohesion.	Adhesion.	Notes on Form, Æstivation, Colour, etc.
Perianth. *Leaves.*				
Calyx. *Sepals.*				
Corolla. *Petals.*				
Stamens. *Filaments. Anthers.*				
Pistil. *Stigmas. Styles. Carpels. Ovary-cells.*				
Fruit. Kind Variety Dehiscence No. of Seeds Description of Seed				

Floral Diagram.

FLOWER SCHEDULES.

Flower of ....

Organ.	No.	Cohesion.	Adhesion.	Notes on Form, Æstivation, Colour, etc.
Perianth. *Leaves.*				
Calyx. *Sepals.*				
Corolla. *Petals.*				
Stamens. *Filaments.* *Anthers.*				
Pistil. *Stigmas.* *Styles.* *Carpels.* *Ovary-cells.*				

Fruit.
Kind ....
Variety ....
Dehiscence ....
No. of Seeds ....
Description of Seed ....

Floral Diagram.

Flower of ....

Organ.	No.	Cohesion.	Adhesion.	Notes on Form, Æstivation, Colour, etc.
Perianth. *Leaves.*				
Calyx. *Sepals.*				
Corolla. *Petals.*				
Stamens. *Filaments.* *Anthers.*				
Pistil. *Stigmas.* *Styles.* *Carpels.* *Ovary-cells.*				

Fruit.
Kind ....
Variety ....
Dehiscence ....
No. of Seeds ....
Description of Seed ....

Floral Diagram.

FLOWER SCHEDULES.

FLOWER OF _____

ORGAN.	No.	COHESION.	ADHESION.	NOTES ON FORM, ÆSTIVATION, COLOUR, ETC.
Perianth. Leaves.				
Calyx. Sepals.				
Corolla. Petals.				
Stamens. Filaments. Anthers.				
Pistil. Stigmas. Styles. Carpels. Ovary-cells.				

FRUIT. Kind
Variety
Dehiscence
No. of Seeds
Description of Seed

FLORAL DIAGRAM.

FLOWER OF _____

ORGAN.	No.	COHESION.	ADHESION.	NOTES ON FORM, ÆSTIVATION, COLOUR, ETC.
Perianth. Leaves.				
Calyx. Sepals.				
Corolla. Petals.				
Stamens. Filaments. Anthers.				
Pistil. Stigmas. Styles. Carpels. Ovary-cells.				

FRUIT. Kind
Variety
Dehiscence
No. of Seeds
Description of Seed

FLORAL DIAGRAM.

FLOWER SCHEDULES.

Flower of ....

Organ.	No.	Cohesion.	Adhesion.	Notes on Form, Æstivation, Colour, etc.
Perianth. Leaves.				
Calyx. Sepals.				
Corolla. Petals.				
Stamens. Filaments. Anthers.				
Pistil. Stigmas. Styles. Carpels. Ovary-cells.				

Fruit.
  Kind
  Variety
  Dehiscence
  No. of Seeds
  Description of Seed

Floral Diagram.

Flower of ....

Organ.	No.	Cohesion.	Adhesion.	Notes on Form, Æstivation, Colour, etc.
Perianth. Leaves.				
Calyx. Sepals.				
Corolla. Petals.				
Stamens. Filaments. Anthers.				
Pistil. Stigmas. Styles. Carpels. Ovary-cells.				

Fruit.
  Kind
  Variety
  Dehiscence
  No. of Seeds
  Description of Seed

Floral Diagram.

FLOWER SCHEDULES.

FLOWER OF ....

ORGAN.	No.	Cohesion.	Adhesion.	Notes on Form, Æstivation, Colour, etc.
Perianth. *Leaves.*				
Calyx. *Sepals.*				
Corolla. *Petals.*				
Stamens. *Filaments. Anthers.*				
Pistil. *Stigmas. Styles. Carpels. Ovary-cells.*				

FRUIT.	Kind	
	Variety	
	Dehiscence	
	No. of Seeds	
	Description of Seed	

FLORAL DIAGRAM.

FLOWER OF ....

ORGAN.	No.	Cohesion.	Adhesion.	Notes on Form, Æstivation, Colour, etc.
Perianth. *Leaves.*				
Calyx. *Sepals.*				
Corolla. *Petals.*				
Stamens. *Filaments. Anthers.*				
Pistil. *Stigmas. Styles. Carpels. Ovary-cells.*				

FRUIT.	Kind	
	Variety	
	Dehiscence	
	No. of Seeds	
	Description of Seed	

FLORAL DIAGRAM.

FLOWER SCHEDULES.

Flower of ....

Organ.	No.	Cohesion.	Adhesion.	Notes on Form, Æstivation, Colour, etc.
Perianth. *Leaves.*				
Calyx. *Sepals.*				
Corolla. *Petals.*				
Stamens. *Filaments. Anthers.*				
Pistil. *Stigmas. Styles. Carpels. Ovary-cells.*				
Fruit.	Kind .... Variety .... Dehiscence .... No. of Seeds .... Description of Seed ...			

Floral Diagram.

Flower of ....

Organ.	No.	Cohesion.	Adhesion.	Notes on Form, Æstivation, Colour, etc.
Perianth. *Leaves.*				
Calyx. *Sepals.*				
Corolla. *Petals.*				
Stamens. *Filaments. Anthers.*				
Pistil. *Stigmas. Styles. Carpels. Ovary-cells.*				
Fruit.	Kind .... Variety .... Dehiscence .... No. of Seeds .... Description of Seed ...			

Floral Diagram.

FLOWER SCHEDULES.

Flower of

Organ.	No.	Cohesion.	Adhesion.	Notes on Form, Æstivation, Colour, etc.
Perianth. Leaves.				
Calyx. Sepals.				
Corolla. Petals.				
Stamens. Filaments. Anthers.				
Pistil. Stigmas Styles. Carpels Ovary-cells.				
Fruit. Kind Variety Dehiscence No. of Seeds Description of Seed				

Floral Diagram.

Flower of

Organ.	No.	Cohesion.	Adhesion.	Notes on Form, Æstivation, Colour, etc.
Perianth. Leaves.				
Calyx. Sepals.				
Corolla. Petals.				
Stamens. Filaments. Anthers.				
Pistil. Stigmas Styles. Carpels Ovary-cells.				
Fruit. Kind Variety Dehiscence No. of Seeds Description of Seed				

Floral Diagram.

FLOWER SCHEDULES.

FLOWER OF..

Organ.	No.	Cohesion.	Adhesion.	Notes on Form, Æstivation, Colour, etc.
Perianth. *Leaves.*				
Calyx. *Sepals.*				
Corolla. *Petals.*				
Stamens. *Filaments. Anthers.*				
Pistil. *Stigmas. Styles. Carpels. Ovary-cells.*				

FRUIT.
- Kind ....
- Variety ....
- Dehiscence ....
- No. of Seeds ....
- Description of Seed ....

FLORAL DIAGRAM.

FLOWER OF..

Organ.	No.	Cohesion.	Adhesion.	Notes on Form, Æstivation, Colour, etc.
Perianth. *Leaves.*				
Calyx. *Sepals.*				
Corolla. *Petals.*				
Stamens. *Filaments. Anthers.*				
Pistil. *Stigmas. Styles. Carpels. Ovary-cells.*				

FRUIT.
- Kind ....
- Variety ....
- Dehiscence ....
- No. of Seeds ....
- Description of Seed ....

FLORAL DIAGRAM.

FLOWER SCHEDULES.

Flower of

Organ.	No.	Cohesion.	Adhesion.	Notes on Form, Æstivation, Colour, etc.
Perianth. *Leaves.*				
Calyx. *Sepals.*				
Corolla. *Petals.*				
Stamens. *Filaments. Anthers.*				
Pistil. *Stigmas. Styles. Carpels. Ovary cells.*				

Fruit. Kind
      Variety
      Dehiscence
      No. of Seeds
      Description of Seed

Floral Diagram.

Flower of

Organ.	No.	Cohesion.	Adhesion.	Notes on Form, Æstivation, Colour, etc.
Perianth. *Leaves.*				
Calyx. *Sepals.*				
Corolla. *Petals.*				
Stamens. *Filaments. Anthers.*				
Pistil. *Stigmas. Styles. Carpels. Ovary cells.*				

Fruit. Kind
      Variety
      Dehiscence
      No. of Seeds
      Description of Seed

Floral Diagram.

FLOWER SCHEDULES.

Flower of ....

Organ.	No.	Cohesion.	Adhesion.	Notes on Form, Æstivation, Colour, etc.
Perianth. *Leaves.*				
Calyx. *Sepals.*				
Corolla. *Petals.*				
Stamens. *Filaments. Anthers.*				
Pistil. *Stigmas. Styles. Carpels. Ovary-cells.*				

Fruit.
- Kind ....
- Variety ....
- Dehiscence ....
- No. of Seeds ....
- Description of Seed ....

Floral Diagram.

Flower of ....

Organ.	No.	Cohesion.	Adhesion.	Notes on Form, Æstivation, Colour, etc.
Perianth. *Leaves.*				
Calyx. *Sepals.*				
Corolla. *Petals.*				
Stamens. *Filaments. Anthers.*				
Pistil. *Stigmas. Styles. Carpels. Ovary-cells.*				

Fruit.
- Kind ....
- Variety ....
- Dehiscence ....
- No. of Seeds ....
- Description of Seed ....

Floral Diagram.

FLOWER SCHEDULES.

Flower of _____

Organ.	No.	Cohesion.	Adhesion.	Notes on Form, Æstivation, Colour, etc.
Perianth. Leaves.				
Calyx. Sepals.				
Corolla. Petals.				
Stamens. Filaments. Anthers.				
Pistil. Stigmas. Styles. Carpels. Ovary-cells.				
Fruit. Kind Variety Dehiscence No. of Seeds Description of Seed				

Floral Diagram.

Flower of _____

Organ.	No.	Cohesion.	Adhesion.	Notes on Form, Æstivation, Colour, etc.
Perianth. Leaves.				
Calyx. Sepals.				
Corolla. Petals.				
Stamens. Filaments. Anthers.				
Pistil. Stigmas. Styles. Carpels. Ovary-cells.				
Fruit. Kind Variety Dehiscence No. of Seeds Description of Seed				

Floral Diagram.

FLOWER SCHEDULES.

Flower of ....

Organ.	No.	Cohesion.	Adhesion.	Notes on Form, Æstivation, Colour, etc.
Perianth. *Leaves.*				
Calyx. *Sepals.*				
Corolla. *Petals.*				
Stamens. *Filaments. Anthers.*				
Pistil. *Stigmas. Styles. Carpels. Ovary-cells.*				

Fruit.
- Kind ....
- Variety ....
- Dehiscence ....
- No. of Seeds ....
- Description of Seed ....

Floral Diagram.

Flower of ....

Organ.	No.	Cohesion.	Adhesion.	Notes on Form, Æstivation, Colour, etc.
Perianth. *Leaves.*				
Calyx. *Sepals.*				
Corolla. *Petals.*				
Stamens. *Filaments. Anthers.*				
Pistil. *Stigmas. Styles. Carpels. Ovary-cells.*				

Fruit.
- Kind ....
- Variety ....
- Dehiscence ....
- No. of Seeds ....
- Description of Seed ....

Floral Diagram.

FLOWER SCHEDULES.

FLOWER OF _____

ORGAN.	No.	COHESION.	ADHESION.	NOTES ON FORM, ÆSTIVATION, COLOUR, ETC.
Perianth. *Leaves.*				
Calyx. *Sepals.*				
Corolla. *Petals.*				
Stamens. *Filaments. Anthers.*				
Pistil. *Stigmas. Styles. Carpels. Ovary-cells.*				
FRUIT. Kind Variety Dehiscence No. of Seeds Description of Seed				

FLORAL DIAGRAM.

FLOWER OF _____

ORGAN.	No.	COHESION.	ADHESION.	NOTES ON FORM, ÆSTIVATION, COLOUR, ETC.
Perianth. *Leaves.*				
Calyx. *Sepals.*				
Corolla. *Petals.*				
Stamens. *Filaments. Anthers.*				
Pistil. *Stigmas. Styles. Carpels. Ovary-cells.*				
FRUIT. Kind Variety Dehiscence No. of Seeds Description of Seed				

FLORAL DIAGRAM.

FLOWER SCHEDULES.

FLOWER OF ....

ORGAN.	No.	COHESION.	ADHESION.	NOTES ON FORM, ÆSTIVATION, COLOUR, ETC.
Perianth. *Leaves.*				
Calyx. *Sepals.*				
Corolla. *Petals.*				
Stamens. *Filaments. Anthers.*				
Pistil. *Stigmas. Styles. Carpels. Ovary-cells.*				

FRUIT.
Kind ....
Variety ....
Dehiscence ....
No. of Seeds ....
Description of Seed ....

FLORAL DIAGRAM.

FLOWER OF ....

ORGAN.	No.	COHESION.	ADHESION.	NOTES ON FORM, ÆSTIVATION, COLOUR, ETC.
Perianth. *Leaves.*				
Calyx. *Sepals.*				
Corolla. *Petals.*				
Stamens. *Filaments. Anthers.*				
Pistil. *Stigmas. Styles. Carpels. Ovary-cells.*				

FRUIT.
Kind ....
Variety ....
Dehiscence ....
No. of Seeds ....
Description of Seed ....

FLORAL DIAGRAM.

FLORAL DIAGRAMS.

Diagram of ....

Diagram of ......

Diagram of ....

Diagram of

Diagram of.

Diagram of.

Diagram of

Diagram of

FLORAL DIAGRAMS.

Diagram of..............

Diagram of

Diagram of..............

Diagram of..............

Diagram of..............

Diagram of..............

Diagram of..............

Diagram of..............

INDEX OF PLANTS.

NO.	NAME OF PLANT.	NO.	NAME OF PLANT.

www.ingramcontent.com/pod-product-compliance
Lightning Source LLC
Chambersburg PA
CBHW030302170426
43202CB00009B/844